W0230488

LANDSCAPE ECOLOGY

LANDSCHAFTSFORSCHUNG UND ÖKOLOGIE

BIOGEOGRAPHICA

Editor-in-Chief

J. SCHMITHÜSEN

Editorial Board

L. BRUNDIN, Stockholm; H. ELLENBERG, Göttingen; J. ILLIES, Schlitz;
H. J. JUSATZ, Heidelberg; C. KOSSWIG, Istanbul; A. W. KÜCHLER, Lawrence;
H. LAMPRECHT, Göttingen; A. MIYAWAKI, Yokohama; W. F. REINIG, Hardt;
S. RUFFO, Verona; H. SICK, Rio de Janeiro; H. SIOLI, Plön; V. SOTCHAVA,
Irkutsk; V. VARESCHI, Caracas; E. M. YATES, London

Secretary
P. MÜLLER, Saarbrücken

VOLUME 16

DR. W. JUNK B.V., PUBLISHERS, THE HAGUE-BOSTON-LONDON 1979

LANDSCAPE ECOLOGY

LANDSCHAFTSFORSCHUNG UND ÖKOLOGIE

Volume in Honour of Prof. Dr. J. Schmithüsen   Festschrift zu Ehren von Prof. Dr. J. Schmithüsen

DR. W. JUNK B.V., PUBLISHERS, THE HAGUE-BOSTON-LONDON 1979

ISBN-13:978-94-009-9621-2   e-ISBN-13:978-94-009-9619-9
DOI: 10.1007/978-94-009-9619-9

© Dr. W. Junk Publishers, The Hague 1979
Softcover reprint of the hardcover 1st edition 1979

Cover design: Max Velthuijs

CONTENTS/INHALT

LIEBER JOSEF SCHMITHÜSEN

Der vorliegende Sammelband der Reihe Biogeographica, die 1972 von Ihnen begründet wurde, sei Ihnen zu Ihrem 70. Geburtstage am 30. Januar 1979 gewidmet. Der Band hat den Titel "Landschaftsforschung und Ökologie" erhalten, er möchte zur Diskussion von Fragen beitragen, mit denen Sie sich in langjähriger Tätigkeit als Forscher und akademischer Lehrer beschäftigt haben, und Sie auf diese Weise ehren.

Ihre Kollegen im Geographischen Institut der Universität des Saarlandes, dem Sie bis zu Ihrer Emeritierung 1977 aktiv angehörten und mit dem Sie seither weiter eng verbunden geblieben sind, hatten Ihnen schon einmal zu Ihrem 65. Geburtstage im Jahre 1974 für Ihr Wirken in Lehre und Forschung gedankt. Die damalige Schrift (Arbeiten aus dem Geographischen Institut der Universität des Saarlandes, Band 18, 1974) enthielt unter dem Titel 'Landschaft und Vegetation' die wesentlichsten Ihrer gesammelten Aufsätze von 1934 bis 1971. Diese Zusammenstellung wollte Ihr wissenschaftliches Werk für sich selbst sprechen lassen und zugleich den Wunsch vieler Freunde und Schüler nach dem Besitz Ihrer kürzeren in Zeitschriften verstreuten Arbeiten erfüllen.

Diesmal wurde der Versuch unternommen, einige Forscher des In- und Auslandes zu Worte kommen zu lassen, die sich intensiver um die Fragen bemüht haben, die immer wieder im Mittelpunkte Ihres wissenschaftlichen Werkes gestanden haben und denen in wörtlicher Übereinstimmung mit dem von uns gewählten Titel das letzte Kapitel Ihrer Grundlagen der Landschaftskunde (Allgemeine Geosynergetik) von 1976 gewidmet ist. In diesem Werke haben Sie die Gedanken und Schlussfolgerungen zusammengefasst, zu denen Sie in jahrzehntelanger Beschäftigung mit den Fragen der geographischen Landschaft und ihrer ökologischen Aspekte gelangt sind.

Dieses Interesse bestimmte schon Ihre Dissertation über den Niederwald des linksrheinischen Schiefergebirges, die 1934 unter Leo Waibel in Bonn abgeschlossen wurde. Am Beispiel einer vom Menschen intensiv genutzten und umgestalteten Waldformation wurden Fragen sowohl der Vegetationsgeographie als auch des von der menschlichen Wirtschaft bestimmten ökologischen Wirkungsgefüges angesprochen, so dass die Arbeit mit Recht ein Beitrag zur Geographie der Kulturlandschaft genannt werden konnte. Nahezu gleichzeitig erwachte auch schon Ihr Interesse an der räumlichen Gliederung Ihrer Arbeitsgebiete, die sich zunächst nach Westen ausdehnten. Nach verschiedenen Vorarbeiten legten Sie 1940 als Habilitationsschrift der Universität Bonn eine Darstellung des Luxemburger Landes vor, in der freilich die Behandlung des städtischen und des industriellen Bereiches noch fehlte. Ihr Habilitationsvortrag, der 1942 in erweiterter Fassung unter dem Titel 'Vegetationsforschung und ökologische Standortslehre in ihrer Bedeutung für die Geographie der Kulturlandschaft' veröffentlicht wurde, umreisst im Grunde bereits alle Gedankengänge und Fragen, die auch in Ihren folgenden Arbeiten immer wieder aufgeworfen werden und bis zu Ihrem Buche von 1976 ständig weiter vertieft worden sind. Ihre Arbeiten im linksrheinischen Schiefergebirge und an der

Mosel, in Luxemburg und Lothringen machten Sie ausserdem zu einem aktiven Mitarbeiter der von Wilhelm Credner begründeten Landwirtschaftsgeographischen Arbeitsgemeinschaft, die in den Jahren vor dem letzten Kriege eine rege Tätigkeit entfaltete und die vor allem durch die von ihr angeregten grossmasstäblichen Flurkartierungen, denen Sie den farbigen Zeichenschlüssel lieferten, der Agrargeographie methodisch neue Wege gewiesen hat. Ihre eigene Kartierung der land- und forstwirtschaftlichen Nutzflächen an der Mittelmosel und im Hunsrück fand 1962 schliesslich Aufnahme in dem von Erich Otremba herausgegebenen Atlas der deutschen Agrarlandschaft.

Schon als junger Dozent begannen Sie, mit der neugegründeten, unter der Leitung Ihres langjährigen Freundes Emil Meynen stehenden Abteilung für Landeskunde beim Reichsamt für Landesaufnahme, der späteren Bundesanstalt für Raumforschung und Landeskunde in Bad Godesberg eng zusammenzuarbeiten. Hier oblag Ihnen insbesondere die Betreuung des grossen Unternehmens der naturräumlichen Gliederung; sie erstreckte sich einmal auf die Geländekartierungen auf der Grundlage der topographischen Übersichtskarte 1:200.000, zu denen Sie selbst 1952 das Blatt Karlsruhe, die Umgebung Ihres Wirkungsfeldes als Hochschullehrer bis 1962, beigesteuert haben; sie betraf aber auch das zusammenfassende Handbuch der naturräumlichen Gliederung der ganzen Bundesrepublik Deutschland, das Sie von 1953 bis 1962 zusammen mit Emil Meynen herausgegeben haben und in dem die Einleitung mit den grundsätzlichen und methodischen Richtlinien aus Ihrer Feder stammt. Die Ergebnisse dieses Werkes, das zunächst nur administrativen und statistischen Zwecken dienen sollte, haben inzwischen für die Raumplanung von Umweltschutz, Grünflächenerhaltung und Naherholung grosses Gewicht gewonnen. Im Rahmen dieser Arbeiten an der Bundesanstalt verdienen auch Ihre Studien über Dürreschäden und Dürre-Empfindlichkeit der mitteleuropäischen Wirtschaftslandschaft nach dem trockenen Sommer 1947 genannt zu werden, die leider nur skizzenhaft publiziert worden sind und die dringend wieder aufgegriffen werden sollten.

Die zunehmende Bewegungsfreiheit nach dem Kriege hat es Ihnen endlich auch ermöglicht, grössere Reisepläne im Auslande zu verwirklichen, die wie für viele Angehörige Ihrer Generation lange Zeit hatten zurückstehen müssen. Unter den mediterranen Ländern, die Sie schon als junger Mann besuchen konnten, und ihren Vegetationsformationen hatte Sie besonders Spanien fasziniert, wohin Sie auch später mehrmals Studentenexkursionen geführt und über das Sie Dissertationen angeregt haben, – ganz abgesehen von Ihrem Zweitwohnsitz auf der Insel Formentera. So lag es nahe, das spanischsprechende Südamerika als überseesisches Arbeitsgebiet zu wählen. Sie begannen 1952 in Chile zu forschen, bereisten in der Folge auch andere Andenländer, und die vegetationsgeographischen Zusammenhänge veranlassten Sie endlich, den zirkumpazifischen Raum näher ins Auge zu fassen und Geländekenntnisse auch in Neuseeland, Japan und Indonesien zu erwerben. Besonders der Vergleich zwischen den südhemisphärischen Gebirgsländern Chile und Neuseeland hat Sie zu vielen interessanten Beobachtungen und Feststellungen angeregt.

Damit war der Grund gelegt, auf dem das Gebäude Ihrer allgemeinen und theoretischen Überlegungen aufgeführt werden konnte. Trotz der Vielseitigkeit der Themen, mit denen Sie sich im Laufe der Zeit befasst haben, lassen sich doch ganz klar einige Schwerpunkte herauskristallisieren, deren Diskussion von Ihnen besonders gefördert worden ist. Es liegt nahe, mit der Vegetationsgeographie zu beginnen, zumal Ihnen Ihre gediegene Beherrschung der gesamten Wissenschaft von der Vegetation auch die volle Anerkennung des Nachbarfaches Botanik eingebracht hat. Erst Ihrer Konzeption ist es gelungen, die Vegetationsgeographie klar gegen die Verbreitungslehre der Geobotanik abzugrenzen und voll in die Geographie zu integrieren. Seit Ihrem Vortrage 1957 auf dem Deutschen Geographentag in Würzburg hat sich auch der Begriff Vegetationsgeographie mit dem von Ihnen gewünschten Inhalt mehr und mehr durchgesetzt. Die bald darauf folgende 'Allgemeine Vegetationsgeographie' im Lehrbuch der Allgemeinen Geographie von Erich Obst, dessen Mitherausgeber Sie inzwischen geworden sind, hat bis heute bereits drei Auflagen erreicht. Nach mehreren Beiträgen zu Problemen der Vegetationskartierung hat dieser Teil Ihrer wissenschaftlichen Arbeit einen vorläufigen krönenden Abschluss im 1976 erschienenen Atlas zur Biogeographie des Bibliographischen Institutes gefunden, einem Werk, an dem auch zahlreiche von Ihnen angeleitete Schüler mitwirken konnten.

Obwohl die historische Dimension der Landschaftsentwicklung schon in vielen Ihrer frühen Arbeiten erkennbar wird, rückt dieser Aspekt geographischer Forschung doch erst relativ spät in den Vordergrund Ihrer Publikationstätigkeit. Mancher, der Sie für einen Vegetationsgeographen mit Neigungen zur Agrargeographie und Landschaftsökologie gehalten hatte, mag erstaunt gewesen sein, als 1970 eine 'Geschichte der Geographischen Wissenschaft von den ersten Anfängen bis zum Ende des 18. Jahrhunderts' aus Ihrer Feder erschien. Wir hoffen auf eine Fortsetzung dieser übersichtlichen Darstellung bis in die Gegenwart. Bei näherem Hinsehen wird dieses Interesse für die Geschichte unseres Faches leicht verständlich. Sie haben Disziplingeschichte nie um ihrer selbst willen betrieben, sondern immer in dem Bestreben, den Ursprung und den Entwicklungsgang von Begriffen, Hypothesen und Theorien in der Geographie aufzuspüren und den geistigen Vorgängern von Ideen zur gebührenden Anerkennung zu verhelfen. Sie haben damit nicht nur wichtige Fakten der Disziplingeschichte aufklären helfen, sondern auch Ihrem eigenen Ideengebäude ein sicheres Fundament gegeben.

Dieses System Ihrer Vorstellungen von dem Wesen und den Aufgaben der Geographie hat selbst schon eine lange Entwicklungsgeschichte hinter sich. Ansätze waren, wie betont, schon in Ihrer Doktorarbeit zu erkennen und lassen sich durch viele Aufsätze zur Vegetationsforschung und naturräumlichen Gliederung verfolgen. Erstmals 1949 haben Sie dann zusammen mit Hans Bobek das Konzept der 'Landschaft im logischen System der Geographie' vorgelegt. Richtungsweisend wurden auch Ihre Beiträge 1959 über 'das System der Geographischen Wissenschaft' in der Festschrift für Theodor Kraus und 1963 und 1964 über den Landschaftsbegriff, den Sie auch auf seine historischen Ursprünge zurückverfolgt haben und dem Sie zuletzt auch in der Entwicklung der europäischen Landschaftsmalerei nachgegangen sind. Dabei ergab es sich fast zwangsläufig, dass auch der von Leo Waibel geprägte Formationsbegriff der Wirtschaftsgeographie, der ja aus dem biotischen Bereiche abgeleitet ist, mit in die Diskussion einbezogen wurde. Sie haben keine Gelegenheit versäumt, um Ihr theoretisches System auf nationalen und internationalen Symposien mit Fachkollegen zu diskutieren. Mit besonders nachdrücklichen Erfolgen ist dies 1965 auf dem Symposium 'Theorie der Geographie' geschehen, das Sie selbst mit Unterstützung der Deutschen Forschungsgemeinschaft in Saarbrücken veranstalteten. Es soll nicht verschwiegen werden, dass Ihre Konzeption der Geographie gerade im engeren Fache zum Teil auf heftige Kritik gestossen ist, insbesondere auf und seit dem Deutschen Geographentage 1969 in Kiel. Diese Angriffe gegen den Landschaftsbegriff müssen im grösseren Rahmen der Auseinandersetzungen um die Einheit der Geographie und um die geographische Länderkunde gesehen werden. Unseres Erachtens hat sich Ihre Konzeption behauptet, wenn auch ein Erfolg Ihrer Bemühungen, eine international verständliche Fachterminologie zu entwickeln, für die Sie 1962 Vorschläge publiziert und die Sie in Ihrem Werke über die 'Allgemeine Geosynergetik' von 1976 konsequent zur Anwendung gebracht haben, wohl noch längere Zeit auf sich warten lassen wird. Die jüngste Entwicklung zeigt ja deutlich, dass der Landschaftsbegriff, falls er von der Geographie nach harten Auseinandersetzungen aufgegeben werden sollte, sofort von einer Reihe anderer Wissenschaftszweige für sich in Anspruch genommen wird. Das, was Sie mit diesem Begriffe meinen, wird sich also auf jeden Fall in der Wissenschaft behaupten, auch wenn die Zielsetzungen und Arbeitsrichtungen der Geographie einem Wandel unterliegen.

Die folgenden Beiträge sind thematisch weit gestreut, sie sollen daher für sich selbst aussagen. Die Beiträge sind jedoch untereinander dadurch verbunden, dass sie mit Blick auf die Person und das Werk von Josef Schmithüsen verfasst worden sind, in der Hoffnung, nicht nur die wissenschaftliche Erkenntnis zu fördern, sondern auch um Ihnen an Ihrem Jubiläumstage eine Freude zu machen.

Carl Rathjens   Paul Müller

SOME AXIOMS OF VEGETATION SCIENCE

V.B. Sotchava

The vegetation science proceeds from certain points of departure – axioms, as every other branch of science does. All the argumentation concerning this sphere of knowledge is based on these axioms.

Some of such axioms are being discussed below, those related in one form or other to the problems of vegetation mapping, map legends elaboration and the analysis of regularities in the spatial structure of vegetation. Not all of these axioms are generally accepted – and then they are suggested to take place of the existing ones or to increase the store of vegetation science conceptions. The majority of them are tested by the geobotanists of the Institute of Geography of Siberia and the Far East, Irkutsk, in the course of their practical work.

The author is pleased by the possibility to publish this short paper in the book dedicated to the jubilee of the eminent landscape-connoisseur and biogeographer Professor I. Schmithüsen, all the more because in our conception there is a synthesis of landscape and biogeographical principles – that stands close to the scientific interests of the hero of the anniversary and partly coincides with his own tendencies.

1. Plant community as an open system

The plant community – usually examined by geobotanists either floristically or ecologically – is a certain type of open systems, first of all, and that is its main feature, and should be accepted as the initial axiom of the vegetation science. The system conception, being applied to the vegetation, envisages the structure of vegetation, its qualitative composition and ecological linkages and interrelations of its components with each other. As every other system, the vegetation functionates in the definite environment, first of all in that geosystem of which this plant community is one of components, and further – the geosystems of higher ranks. The different ties inside the system give rise to the dynamic processes in the community – both reversible (seasonal rythms among them) and irreversible ones. They are the basis of vegetational systems self-regulation and appear as the motive force of vegetation development – the exchange of substances with the environment (metabolism). As a matter of fact the system concept facilitates for us the physical and chemical explanations of vital processes in the community and their dependence on geographic factors. It determinates the stabilizing role of vegetation in biogeocenoses, their self-regulation and so-called phytocenotic relations.

The open systems are characterized by the constant entrance of substances from outside. Inside the system these substances undergo some transformations, that results in an emergence of the products of great complexity; L. Bertalanffy (1972) named this process metabolism, and he suggested its existance not only in organisms (that had been practicized before him as well), but in all open systems. In certain conditions the open systems show the phenomenon of dynamic equilibrium, that is inherent in a number of plant communities, though as a dynamic feature, not as the evolutional one. The general theory of systems ensures the vegetation science an alliance with cybernetics – the science of ties and regulations. It suggests the cooperation with other natural sciences, whose parameters naturally associate themselves with the graphic and mathematical models of plant community. It is very essential, if to have in mind the modern tendencies towards the intersciences investigations.

The concept of phytocenosystems enables us to combine the spatial principle with the functional one, while in the case of pure floristic treatment of the plant association the latter is often neglected. If we realize the community as a system, we assume some degree of freedom in the connections amongst its components. In particular the ecological and other interdependences are not absolute, but have some degree of variability, i.g. the vegetation is ranged within certain amplitudes of warmth, humidity and other requirements, that promotes the conservation of community structure under the seasonal and yearly fluctuations in environment.

The realization of plant community as a system creates the necessary prerequisites for the mathematical and statistical analysis of components' interdependances, between the composition and the ecological factors in particular. Besides, the system analysis opens certain perspectives of feedbacks, which ensure, although not always, the stability of the system. On the whole, the internal contradictions of the system give rise to its permanent motion, that is the dialectic process.

The system organization of plant community can be viewed as the point of departure in our idea of this community, and this would help us to organize and to regulate our scientific knowledge in vegetation. In the sense of Kuhn (1962) it is the most important paradigm of biogeography.

2. Dimensionality of vegetation

The features of vegetational cover are determined by its dimensionality first of all. Thus we must to primarily indicate – on what area the vegetation we deal with is situated: planetary, regional, subregional or topological. The above-named categories coincide with the dimensions of landscapes and are distinguished by regularities of their own and their existence also appears as an axiom of vegetation cover geography.

Every dimensionality has its peculiarities determined by the volume and mass of phytosphere contained in it. The relationship between biota and environment manifests itself in its own way within the limits of every dimensionality. The dimensionalities of the living mantle of the earth do not only differ in quantity, but every of them represents a special quality. The most universal distinctive sign of dimensity is the mode of metabolism between the plant cover and the environment, i.g. the peculiarities of functioning. The metabolism of topological order occurs against the background of the subregional one, and the latter is but a part of regional metabolism, which in his turn depends on the planetary metabolism. Certainly, within the dimensionality the plant cover metabolism has its local peculiarities, that is one of the main criteria of the geobotanical mapping. Every dimensionality is represented by several taxa of geobotanical regions. However, the regionalization is based on its own axiom.

The quantitative and qualitative aspects of metabolism taking place in definite space frames are peculiar to every dimensionality. The unequal significance of geometrical dimensions of practically existing phytocenosystems belonging to different dimensionalities is characteristic too. That is why dealing with vegetational cover one prefers to use the vectorial measurement units in general and for length measurements in particular. This is valid for mass either, where two units are possible: the mass index as the degree of inertia and the mass unit as the measure of substance quantity — both are essential for the vegetation science.

The analysis of vegetation dimensionalities has yet to be elaborated in order to ascertain the space units relations to each other, that is useful for the knowledge of their essence. It is natural in this case to divide the plant cover into the very broad classes of dimensionality, which already help us to orientate ourselves in the orders of regularities peculiar to plant communities.

The working out of the vegetation dimensionality problem is at its very beginning now, on the stage of the question statement. Perhaps, the specific measurement units and specific mathematical method will be needed here. In due course the more precise criteria for the recognition of plant cover units, as representatives of one or other dimensionality category, will be found. That will draw in this problem more scientists and make the division of vegetation according to its dimensionalities generally accepted. The problem of the dimensionality of investigated objects was already suggested in biogeography (Müller, 1977) and in landscape science (Haase, 1973).

3. Phytocenomeres and phytocenochores

One of the basic concepts of the vegetation science rests on the following axiom. There are two principles acting in the phytogeosphere: homogenization of vegetation and its differentiation. As a consequence of these processes all the plant communities may be subdivided in two categories: phytocenomeres and phytocenochores. The first category is characterized by homogeneity, differently expressed depending on the rank of a phytocenomere, the second one – by its heterogeneity. The homogenization and the differentiation processes accompany the evolution of plant cover including its present dynamics. This is one of the principle axioms of the vegetation science. Phytocenomeres and phytocenochores are interconnected, but they show quite different qualities. They represent the parts of the single but double-lined classification of the plant cover (Sotchava, 1972).

The differentation processes are favoured by different kinds of plant cover disturbance occurring both spontaneous in the process of its evolution and under the direct and indirect influence of man.

The periods, when development of plant communities proceeds on its own, result in the homogenization: the environment is affected by the community for a long time and the stabilizing processes dominate. The tendency of communities to homogenization is different depending on their capacity to influence the environment, as well as intensity of interactions among the species which constitute them.

The concept of phytocenomere is based on the notion of the homogeneous area of community in the nature. The elementary phytocenomere is a minimal homogeneous space giving room for all components of the given system.

The least taxon of phytocenomere is the association and the phytocenoses is a single representative of association on the earth surface. The similar phytocenomeres are united in an association. The associations are generalized in consecutive order, according to the principle of generalization, in groups and classes, then formations and further in larger taxa, which compose in total the classification line of phytocenomeres parallel to one of phytocenochores.

The differentiation of the plant cover is a historical process, on one hand. The evolution of vegetation is accompanied with the increase of its diversity. And the influence of man also contributes to the differentiation of vegetation. The interelation of the two forms of plant cover's existance – in the types of homogeneous and heterogeneous subdivisions, phytocenomeres and phytocenochores – is important. These subdivisions are expressed on all levels of dimensionality of vegetation cover, but they are especially distinctive in the limits of the topological dimensionality; however on maps they are well recognized on the regional, subregional and planetary levels. The mapping is the principle way for the recognition of phytocenochores in the nature.

The idea of phytocenomeres and phytocenochores has something in common with the Prof. Schmithüsen's (Schmithüsen, Netzel, 1963) concept of synergetics and choretics, which was advanced still in 1962, and is of a great importance for the vegetation science.

4. Hierarchicity of plant communities

With the appearance of general system theory one commences to apply the principle of hierarchy to the study of different system formations, the nature formations among them. Nowadays it begins to penetrate into the geobotany either, but it is already clear, that here it is based on one of the main axioms of the vegetation science.

The ordinary understanding of system hierarchicity is expressed in their division into relatively separated subsystems, which are subordinated to each other. There are two peculiarities we meet in the hierarchicity of the phytocenosystems: their existence in different environment conditions and their age diversity that results in the simultaneous existence of several hierarchical rows of vegetation. In the latter case the macro- and micro-approaches are possible. The macro-approach allows a great number of hierarchical rows, while micro-approach decreases them according to the degree of generalization adopted. It is possible to construct the hierarchical rows of phytocenomeres starting from the three suites (sequences) of vegetation types (north extra-tropical suite, tropical and south extra-tropical ones). Further every suite has to be divided into types of vegetation, the latter — into phratries, which in their turn, into classes and groups of formations etc. Doing so one should keep to the territorial and structural principles and also the evolutionary-historical to some degree, since if any taxon stands on higher level, it has to be older, as a rule.

The hierarchicity is the basis of phytocenosystems' systematization first of all. Also it helps in the elaboration of measures for nature protection and nature reorganisation. The hierarchicity of phytocenochores is well known (microphytocenochore, macrophytocenochore, region, district etc.) and is used in every attempt of regionalization.

The concept of hierarchicity follows from the realization of the plant cover as a system. This concept is of special importance in mapping for it assures the systematic organization of legends. The description of geosystems in a country must be accomplished in the hierarchical sequence — beginning with the large-scale subdivision towards to the smaller and subordinate ones. The primary attention should be paid to the highest members of a hierarchy, because they always are determined by general and significant characteristics. In the hierarchical row of phytocenochores the higher gradations appear as the environment with respect to the lower ones. The hierarchy of the phytocenosystems reflects their evolution to some extent, that is of special importance in the case one analyses some of their components. The idea of phytocenosystems' hierarchy together with the notion of their dimensionality suggests to us an impression about the hierarchy of metabolism; the latter increasing significally from the lower towards the higher levels of the hierarchical row. So the taxonomic scale of phytocenosystems — that is not only the table of their ranks, but it shows the scales of their material-energetic activity as well.

The axiom of the hierarchical subordination of vegetation subdivisions has the comprehensive significance.

5. Evolution and dynamics of the phytocenosphere

As was mentioned above, the vegetation cover is a different-aged phenomenon. Together with the geographical shell of the earth it has passed a long way of development. The geological strata fully lacking of life residue are unknown. The life changed continuously and this long row of changes presents the evolutionary process of vegetation development basically irreversible and characterized by the increase of number of its components and the complication of its structure as a whole (if to consider it globally).

Unlike these evolutionary changes of vegetation, the dynamical transformations peculiar to it are reversal or nearly reversal as a rule. The evolutionary and the dynamical transformations are interrelated and interconditioned, however they have to differ from each other. The dynamics and the evolution are the phenomena of different scale. Although we grasp the evolution as is it were timed to certain geological periods, practically it is a permanent process of plant communities development, their phytocenogenesis. This process results from the fact that the changes which arise of dynamical phenomena are often not fully reversible. So in the long run there accumulate some minute features of evolutionary significance.

But the scales of evolution and ones of the current dynamical changes are not quantitatively comparable. That is why we have right to take as an axiom the fact that the vegetation undergoes the two types of changes — evolutionary and dynamic ones — and to distinguish between the dynamics of vegetation and its evolution, although in some cases one may view the dynamics as the factor of evolution. However, infrequently the dynamics plays the reverse, stabilizing role, that leads in the end to the conservation of the initial type. This is the stabilizing dynamics convincingly suggested by J. Schmithüsen (1968). The problems of dynamics are fully in the competence of geobotanists, while the principles of evolution (philocenogenesis) are developed by them together with paleogeographers and evolutionary botanists.

The subdivision of the developmental processes in plant communities into evolutionary and dynamic ones is based on the following axiom. A certain plant community in a certain period may be characterized both by the invariant condition and by numerous variable conditions. An invariant, which we will discuss apart, — that is what remains unchanged under certain transformations, and the changeable states — its derivatives.

6. Lower subdivisions of the phytocenosphere

As was mentioned above, the axioms of vegetation science include the ambiguous concept of the lower subdivision of vegetation cover. This concept should be considered in four aspects. One of them: the phytocenosis is a solitary expression of a plant association. The second one: separate links of the plant communities series, which were called at one time by F. Clements 'serulas', but more often have the name of serial plant communities. The third one: the area of primary continium, which in fact includes two or more phytocenoses; it is sometimes called the topographic continium (i.e. the part of continium expressed as a plot of certain area). And at last: the fourth lower subdivision of phytosphere we should consider the spatial combinations, the primary agglomeration of discrete phytocenoses (or 'serial phytocenoses' of sections of continium), i.e. an elementary phytoceno-chore.

In the past the main attention of phytocenologists was drawn to the plant association (to generalize it into larger taxa or to analyse its 'content': synusia, strata, socies and oth.). It was the approach of vegetation science leaders (Suka-chev, 1954; Braun-Blanquet, 1964). In a general way it was the unfolding of the decisions of Brusseles Botanical Congress (1910).

Of all abovesaid we can formulate the discussed axiom of vegetation science as follows: the lower taxonomy of phytosphere has four aspects of the lower (most fractional) vegetation state. They reflect: 1. the morphology of homogeneous discrete plant communities, 2. their dynamic state, 3. the elementary spatial agglomeration of phytocenoses, 4. the continium of neighbouring phytocenoses, where it takes place.

What for the phytocenosis, the discrete phenomenon to a certain degree, its generalization into the higher taxa is nearly beyond doubts, although it is prac-tizised by different authors with some deviations. The similar phytocenochores are united into microphytocenochores, macrophytocenochores, regions, districts etc. also according to the steady principle, although these subdivisions are called differently. In the double-row classification of vegetation proposed by us phyto-cenomeres and phytocenochores are classified according to their interconnections.

The separate sections of continium in the nature are presented by interpene-trating phytocenomeres, belonging to the definite associations, groups or classes of associations. In accordance with this they may be placed in the legend as an independent unit subordinated to one or other group or class of associations. The serial associations are generalized into series, the final links of which are repre-sented by climax associations or the similar ones. In the legend they belong to the corresponding maternal association. We have in mind the serial (successional) row in its pure state, that is called by Montpellier ecologists (Godron & Poissonet, 1973) sequentia. On maps (for instance, one of France vegetation, scale 1 : 200.000) the series of plant communities are shown in broader interpretation (Gaussen, 1948). They are to be understood or perceived as a complex of serial associations, or groups of them, having some dominants in common as their maternal body.

Also here is to be mentioned the unification of concrete minute subdivisions of vegetation into sigma-associations, accepted in some Western countries (Beguin Cl. & Hegg O., 1975, Tüxen, 1977), i.g. the associations in the sense of Braun-Blan-quet and his followers.

7. Invariants of the plant communities and their variable conditions

One of the cardinal axioms of the vegetation science reads that every subdivision of plant cover should be conceived abstractly, in an invariant aspect, and at the same time it is characterized by many concrete variables of its condition. The latter are numerous on the globe, nearly they cannot be recorded. Such variables emerge both spontaneously and under the human influence and presently the latter way increased immensely.

Those features of the vegetation cover remaining unaltered under definite current transformations belong to the invariant. However the transformations themselves belong to the vegetation cover dynamics in the broad sense – to the current and reversible change of matter, which is rather changeable in its appearances. The concept of invariability in vegetation science is adequate to some extent to the analoguous concept in crystallography and even in mathematics, although the latter operates with speculative values.

Alongside with this analogy, the concept of invariant, if applied to the vegetation science, gains some peculiarities and is greatly (but not fully) determinated by the invariant state of geographical environment. The vegetation is a component of the geosystem, the component most changeable and sensible to the external factors influence, whereas the rest of nature complex components may remain nearly unchanged.

Before we discuss the features of the invariant as it appears in the vegetation science, let us note its importance both in theoretic and practical respects.

First of all, we should mention the necessity of this concept in the composition of the system of the vegetation cover taxa, keeping in mind that in the nature this cover is represented by a lot of variables verging the state of chaos. Many problems of dynamics as well as the construction of succession rows of plant communities and their various series need the concept of invariant. Also we need it when analysing the ecological potential of a land and the dynamical tendencies of its vegetation.

By the way, all the abovesaid is very important for the prognosis of vegetation changes in the time of technical influence on the geographical shell of the earth. While constructing our geobotanical prognoses we must proceed of the invariant with the view to favour those variable changes of plant communities, which conform with our economic or aesthetic interests. The invariant of the vegetation is not at all fixed, it changes too, that signifies the evolution process in plant communities going slowly but permanently alongside with the violent dynamic processes.

We see that in spite of somewhat problematic character of the invariant concept in vegetation science, we have to do our best in order to determine the invariants of vegetation taxa with the possible degree of confidence.

The specifity of invariant concept conformably to the vegetation cover is determined, first of all, by its dependance of the invariant of the geosystem, which contains this plant cover. Both these concepts are interconditioned to a certain extent. We can speak of the invariant of a type of plant communities in the case this vegetation has not been destroyed, but just transformed by external forces. The complete reversibility is not required here, since some floristic ingredients may remain unrestored. The invariant vegetation is an abstract conception to a certain degree, answering the main features of the land ecological potential. It stands most close to the conception of potential vegetation or the plesioclimax, if to conceive the latter without the possible variable conditions. The more concrete notion of geosystem invariants has been given by us in another place (Sochava, 1978). Essentially the invariant vegetation may be viewed as the vegetation on the geosystem invariant. In this case the two notions would draw together in the sense that the vegetation invariant is nothing but a component of the environment type invariant.

The structure of the plant community is the invariant aspect of the system (in our case — phytocenosystem). Thus, both these conceptions (the invariant and the plant community structure) can be elaborated simultaneously, that imparts to the problem a certain material basis. On the whole we have: a system — an invariant — a structure, such is the equation of the notions hidden under the terms. On this way we approach to the solution of the problem. Unfortunately, scientists realize the structure of vegetation in rather various manners. In order to meet the demands of invariant, the structure must include its dynamic tendencies, for the variable of a forest community may be, in certain cases, a grassland. That is why the characteristic of the invariant should include its dynamic possibilities: the ability to form variable states, that is generally depending on the ecology and geographical position and has to be reflected in the classification of vegetation.

All this should be represented in the graphic model, showing the relations inside the phytosystem invariant. The graph (graphic models) method is of great importance in the solution of this problem, since the graph represents the matter basis of the corresponding investigations. The abiotic and biotic linkages existing inside the invariant are to be fixed best on the graph. That is why we must give a broad way to the modelling (both graphic and mathematical ones) in the vegetation science. The creation of models of plant communities invariants, as well as models of their variables, appears now as an urgent task of the phytocenology. The models are especially important for variables (variable conditions), because on models we can remove many of insignificant features and connections, which may be numerous in such cases. The models of the phytocenoses are of the operative-practical significance either, in particular if we deal with vegetation protection and biosphere defence.

8. Geobotanical fields

The concept of the geobotanical field ensues of the axiom that any geobotanical regularity is restricted, as a rule, by a definite area, that is what is called the geobotanical field. Every geobotanical regularity has its own radius of activity (the type of zonality and altitudinal belt subdivision, the bonding with certain types of soil and rock and many other indicative properties of vegetation). In essence, the very area of the association or the formation represents a geobotanical field. The borders of the geobotanical fields do not coincide with each other, as a rule, their essence being based on a certain axiom, which also lies at the basis of the regional paradigm in the vegetational science. The geobotanical fields, whose first examples I gave as far back as in 1948, can possess various sizes, ranging from rather significant ones, as the whole West-Siberian Plain having the type of plant association zonality of its own, and to a small area of a topological dimensionality, where also its peculiarities of structure and distribution of plant communities can be traced. We assign the name of geobotanical fields to the areas, where the definite structural characteristic is not expressed everywhere, but here and there, depending on the actions of local physico-geographical and ecological factors (for instance, the peculiarities and space alternation of plant communities in termokarst depressions (alases) of Central Yakutia or the area of the association strictly depending on the definite chemical features of soil or any other ecological conditions corresponding to this association only). Thus, we may consider the geobotanical field (that has place in the nature) in all four dimensions – planetary, regional, subregional and topological. Although the geobotanical fields are the areas circumscribing the strictly definite regularities, they may overlap each other. Every space regularity has the field of its own, even if the fields of two or more regularities coincide with one another in the space. It is important to realize that the field is not only a geometrical space, but also the definite ecological characters manifesting themselves in this space.

So the idea of the geobotanical field gives us some additional opportunities for measurement and analysis of the space phenomena in the vegetational science. It gives the new impulses for geobotanical mapping and secures the perspectives of the usage of mathematical methods in the geography of vegetation cover. If the enrichment of geobotanical mapping principles and the mathematization of plant cover geography is a matter of future (although the nearest future), the utilization of geobotanical field methods in the indicative geobotany is a reality at present. The geobotanical field is founded on the definite characteristics of the environment, the same, in many cases, which we seek with the aid of more available observations of vegetation cover structure. Sometimes the geobotanical field coincides in its territory with the geobotanical region. However they are the basically different concepts and the notion of the geobotanical field is by no means equivalent to one of the geobotanical region. The field concept deals only with spatial limits of a single geobotanical phenomenon, while the subdivision into regions aims to outline the distributional sphere of several phenomena, different as they are, but in some ways connected with each other.

In the process of geobotanical mapping we settle out territories, whose vegetation although represented by different structural types, is a unity as a whole. We hold this rule whatever ranks of taxa we try to settle. The outlining of geobotanical fields is the operation of a different kind. In the process of the regionalizing we unite some heterogeneous phytocenotic types, which represent a certain unity on the corresponding earth area. The territories of different ranks gained by us in the process of regionalization are hierarchically interconnected, and may be considered as systems (phytocenochores). Another approach we have when determining the limits of the geobotanical fields, which are not the systems and are interrelated not with each other, but only with those regularities whose areas they include.

9. Conclusion

We have discussed the main axioms of vegetation cover science. They descend from the traditional ideas of phytocenology. The axiom of invariability and of variable conditions of vegetation, as well as all abovesaid on its dynamics, has a particular significance in the systematization of the all plant life diversity. Its importance grows with the increase of variable plant communities number resulting from the man's influence on the nature. Concept of the plant community – as the fundamental type of open systems (phytocenosystems) let us to realize many botanico-geographical regularities in the different, new light. The treatment of the phytocenosystem as a sub-system of the geosystem permits us to include the vegetation science into the Earth science, that is proved by the interpretation of separate vegetation cover axioms meaning.

The axiom of geobotanical fields, if to confront it with the concept of regionalization, opens the broad perspectives for the study of botanical geography objects. This axiom stresses that the regionalisation is the choice of a series a heterogeneous entities rather then areas of the separate geobotanical regularity.

The axiom of the lower phytocenotic units, which treats them in different qualities, brings to conformity the ideas of the discretion and of the continuity of the lower vegetation units, and also the serial phytocenoses and elementary phytocenochores.

The majority of axioms have their analogues in the landscape sphere conformably to geosystems, that is probably connected with the fact that both the regularities existing in vegetation science and in geosystem science are controlled by the generally similar laws of the Earth science.

Некоторые аксиомы учения о растительности (резюме)

Автор обсуждает главнейшие аксиомы учения о растительности, имеющие то или иное отношение к ее картированию и анализу пространственных закономерностей растительного покрова.

Исходной аксиомой признается, что растительное сообщество есть открытая система в понимании Л. фон Берталанфи. Это позволяет совместить пространственное и функциональное его изучение.

Второй аксиомой является представление о размерности растительности.

Устанавливаются четыре категории размерности: 1 – планетарная, 2 – региональная, 3 – субрегиональная и 4 – топологичечкая.

Важной аксиомой является признание действия в фитосфере двух начал: гомогенизации и дифференциации растительности. Отсюда деление растительных сообществ на две группы: фитоценомеров и фитоценохор.

В основе представления о фитоценомерах лежит понятие о гомогенном природном ареале растительности. Фитоценохоры – комплекс фитоценомеров, образующий гетерогенную целостность разных уровней (микрофитоценохоры, макрофитоценохоры, район, округ и т.д.). Существенна взаимосвязь обеих форм существования растительности и единая их двухрядная классификация (Сочава, 1972, 1978).

Следующая аксиома гласит: подразделения растительного покрова образуют последовательную иерархию таксономических единиц.

Понятие иерархии вытекает из представления о растительном сообществе как о системе. Иерархическая таксономическая система фитоценозов – это не только табель о рангах, но и соотношение масштабов их иерархической активности.

Развитие биосферы подчиняется особой аксиоме, согласно которой надо различать два взаимосвязанных процесса: эволюции и динамики фитоценосистем. Эволюция – изменение растительности в масштабах ее развития на Земле, динамика – текущие модификации растительных сообшесб, происходящие спонтанно, а в последнее время – преимущественно под влиянием человека. Эти текущие изменения являются фактором эволюции, но еще не самой эволюцией.

Можно считать аксиомой, что низовые подразделения фитосферы (элементарные ее таксоны) имеют четыре аспекта: 1 – фитоценозы, которые обобщаются в ассоциации, 2 – серийные группировки (серулы), в совокупности составляющие сукцессионный ряд, 3 – элементарный участок континуума, если он имеет место, 4 – элементарная фитоценохора – первичная агрегация фитоценозов.

Кардинальной аксиомой учения о растительности является положение, что каждое подразделение растительного покрова должно мыслиться отвлеченно в инвариантном аспекте, но наряду с этим оно характеризуется многими конкретными переменными состояниями. Инвариантными являются те свойства растительного сообщества, которые остаются неизменными при определенных преобразованиях. Понятие об инварианте помогает упорядочить тот 'хаос' переменных состояний растительных систем, котор-который катастрофически возрастает под влиянием человека.

Последняя аксиома касается понятия о геоботаническом поле, под которым предлагается понимать площадь той или иной закономерности географии растительного покрова.

Большинство аксиом имеет свои аналоги в ландшафтной сфере применительно к геосистемам, что связано с тем, что закономерности учения о растительности и учения о геосистемах управляются сходными законами наук о Земле.

16

Zusammenfassung

Der Verfasser behandelt die Hauptaxiome der Lehre über die Vegetation, die ihre Kartierung und Einschätzung der dimensionalen Gesetzmässigkeiten der Pflanzendecke beinhaltet.

Als Ausgangsaxiom wird folgendes anerkannt, dass die Pflanzengesellschaft ein offenes System nach L. van Bertallanffi darstellt. Damit lässt sich ihre dimensionale und funktionelle Untersuchung vereinbaren.

Als zweites Axiom dient die Darstellung über die Dimension der Vegetation. Es sind 4 Dimensionsarten zu unterscheiden: 1 – planetare, 2 – regionale, 3 – subregionale, 4 – topologische.

Es existieren als wichtige Grundsätze zwei in der Phytosphäre einwirkende Prinzipien: Homogenisierung und Differenzierung der Vegetation anzuerkennen. Damit ist die Einteilung der Pflanzengesellschaften in zwei Gruppen begründet: Phytozönomeren und Phytozönochoren.

Der Darstellung über die Phytozönomeren liegt der Begriff vom gleichartigen natürlichen Vegetationsareal zu Grunde. Phytozönochoren sind ein Komplex von Phytozönomeren, die eine heterogene Integrität verschiedener Stufen (Mikrophytozönochoren, Makrophytozönochoren, Distrikten, Bezirken, Regionen usw) bilden. Wichtig ist der Zusammenhang der beiden Existenzformen der Vegetation und ihre einheitliche Zweireihige Klassifikation (Sotchava, 1972, 1978).

Der nächste Grundsatz lautet: die Unterteilungen der Pflanzendecke bilden eine konsequente Hierarchie der Syntaxonen. Bei der Auffassung über die Hierarchie geht man von der Darstellung der Pflanzengesellschaft als System aus. Das hierarchische taxonometrische System der Phytozönosen ist nicht nur eine Rangtabelle, sondern auch eine Korrelation im Rahmen ihrer hierarchischen Aktivität.

Die Entwicklung der Biosphäre ist dem Sonderaxiom unterstellt, nach dem man zwei voneinanderabhängenden Vorgänge: Evolution und Dynamik der Phytozönosysteme, unterscheiden soll. Die Evolution ist eine Veränderung der Vegetation im Rahmen ihrer Entwicklung auf der Erde; die Dynamik beinhaltet die laufenden Modifikationen der Pflanzengemeinschaften, die spontan, in der jüngsten Zeit aber hauptsächlich unter dem Einfluss des Menschen entstehen. Diese ablaufenden Modifikationen sind als Faktor einer Evolution, keinenfalls aber als die Evolution selbst zu betrachten.

Man darf für ein Axiom folgendes halten, dass die elementaren Unterteilungen der Phytosphäre (ihre elementare Syntaxone) 4 Aspekte haben: 1) die eine Assoziation bildenden Phytozönosen, 2) die im Ganzen aus einer sukzessiven Reihe bestehenden Serien – Gruppierungen (Serulen), 3) die elementare Teilstrecke des Kontinuums, wenn es überhaupt vorhanden ist, 4) die elementare Phytozönochora, die sogenannte primäre Aggregation der Phytozönosen.

Das Hauptaxiom der Lehre über die Vegetation ist ein Leitsatz, dass sich jede Gliederung der Pflanzendecke abstrakt in einem invarianten Aspekt denken lassen, aber demzufolge wird er durch mehrere konkrete variable Verhältnisse gekennzeichnet. Unveränderlich sind solche Eigenschaften der Pflanzendecke, die bei den bestimmten Modifikationen konstant bleiben. Der Begriff über eine Invariante hilft dieses Chaos der variabelen Verhältnisse der Vegetationssysteme einregeln. Dieser Wirwar nimmt durch Menscheneinfluss katastrophal zu.

Im letzten Axiom handelt es sich um einen Begriff des geobotanischen Feldes, unter dem man vorschlägt, den Wirkungsbereich jener oder einer anderen Gesetzmässigkeiten der Geographie der Pflanzendecke aufzufassen.

Die meisten Axiome haben ihre Analoge in der Landschaftssphäre hinsichtlich der Geosysteme, das heisst, dass die Gesetzmässigkeiten der Lehre über die Vegetation und der Lehre über Geosysteme durch identische Gesetze der Wissenschaften von der Erde gesteuert werden.

References

Beguin, Cl. & Hegg, O. 1975. Quelques associations d'associations (Sigma-associations) sur les anticlinaux jurassiens recouverts d'une vegetation naturelle potentielle. (Essai d'analyse scientifique du paysage). *Documents phytosoc.*, 6-14, 9-18, Lille.

Bertalanffy, L. von. 1972. General System Theory. 3-d ed. N.Y., 289 p.

Braun-Blanquet, I. 1964. Pflanzensoziologie. 3. Auflage. Wien-N.Y., 865S.

Gaussen H. 1948. Carte de la vegetation de la France, La feuille 78, Toulouse, (carte).

Godron, M. & Poissonet, J. 1972. Quatre thémes complementaires pour la cartographie de la végétation et du milieu (sequence de végétation, vitesse de cicatrisation de la végétation, diversité du paysage, sensibilité de la végétation), *Bull. Soc. Land.Geogr.*, t.6, fasc. 3, p. 329-356.

Haase, G. 1973. Zur Ausgliederung von Raumeinheiten der chorischen und der regionischen Dimension – dargestellt an Beispielen aus der Bodengeographie. *'Peterm. Geograph. Mitt.'*, 117, 2, S. 81-90.

Kuhn, T.S. 1960. The structure of scientific revolutions. Chicago, 286 p.

Müller, P. 1977. Tiergeographie. Stuttgart, 268 S.

Neef, E. 1967. Die theoretischen Grundlagen der Landschaftslehre. Gotha/Leipzig, 162 S.

Schmithüsen, J. 1968. Allgemeine Vegetationsgeographie. 3. Auflage, Berlin, 463 S.

Schmithüsen, J. & Netzel, E. 1963. Vorschläge zu einer internationalen Terminologie geographischer Begriffe auf der Grundlage des geographischen Synergismus. In: Geographisches Taschenbuch, 1962/1963, Wiesbaden, S. 283-286.

Сочава В.Б. 1972. Классификация растительности как иерархия динамических систем. Вкн.: Геоботаническое картографирование, стр. 3-17.

Sochava, V.B. 1972. Classification of vegetation as a hierarchy of dynamic systems. In: Geobotanical Mapping, p. 3-17.

Сочава В.Б. 1978. Введение в учение о геосистемах. Новосибирск, 318 с.

Sotchava, V.B. 1978. Introduction into Geosystem science, Novosibirsk, 318 p.

Tüxen, R. 1977. Zur Homogenität van Sigmaassoziationen, ihrer syntaxonomischen Ordnung und ihrer Verwendung in der Vegetationskartierung. *Documents phytosocial*, N.S., vol. I, Lille, S. 321-327.

Address of the Author:
Prof. Dr. V.B. Sotchava
Institute of Geography of Siberia and
the Far East SB AS USSR
Institutskaya 1, Irkutsk 33, USSR

LES RELATIONS BIOGEOGRAPHIQUES DES ALPES AVEC LES CHAÎNES CALCAIRES PERIPHERIQUES (JURA, APENNIN, DINARIDES)

P. Ozenda

Malgré sa diversité et sa complexité, la végétation de la chaîne alpine est beaucoup mieux connue que celle des autres grands massifs. Sa position centrale en Europe, sa richesse floristique et ses ressources naturelles ont suscité un très grand nombre de travaux, surtout dans sa partie médiane. Les Alpes ont été le berceau, ou du moins le banc d'essai, de plusieurs écoles phytogéographiques; plus récemment, elles ont fait l'objet d'une grande partie de l'effort accompli en cartographie de la végétation, entraînant une représentation des connaissances plus complète et plus équilibrée que dans les autres chaînes. Il serait, certes, excessif de penser que l'on peut interpréter systématiquement la végétation de ces autres chaînes à partir de celle des Alpes; mais on peut admettre que le modèle alpin constitue une approche intéressante pour l'étude d'une grande partie des montagnes de l'Europe (Ozenda, 1975, 1976, 1978), et peut-être de celles de l'Asie occidentale (Dobremez, Ozenda 1976 et de l'Amérique du Nord.

Dans un essai de synthèse phytogéographique des Alpes sud-occidentales (Ozenda 1966), concernant les Alpes maritimes, la Haute-Provence, le Dauphiné et une partie du Piémont, j'ai été conduit à effectuer de constantes comparaisons avec les autres parties de la chaîne, notamment avec les Alpes sud-orientales, et à proposer finalement pour l'ensemble de l'Arc Alpin une division en secteurs (fig. 1), qui s'est trouvée depuis en bon accord avec les travaux ultérieurs de différents auteurs et qui faisait apparaître d'étroites affinités entre les Préalpes et les chaînes plus extérieures. Ces relations étaient assez claires pour pouvoir proposer de réunir en un seul secteur, dit 'delphino-jurassien', les Préalpes calcaires nord-occidentales et le Jura du Sud. D'autre part, à l'extrémité sud-occidentale de la chaîne alpine, les caractères particuliers que j'avais dégagés de longues recherches sur les Alpes maritimes et ligures m'avaient amené à formuler l'hypothèse de l'individualité d'un secteur préligure assimilable, du fait de ses affinités avec les Préalpes gardésanes et illyriques, à 'un fragment d'Alpes orientales accolé aux Alpes occidentales'. Des recherches poursuivies dans cette voie conduisaient peu après (Barbero, Bono et Ozenda, 1970) à suggérer que ce secteur préligure 'pourrait être la terminaison occidentale d'un ensemble couvrant les parties méditerranéennes des Balkans et de l'Italie': c'est cette hypothèse élargie qui sera développée dans les deuxième et troisième parties du présent exposé.

Le moment paraît venu de tenter de réunir en une vue globale les affinités biogéographiques des Alpes avec leurs trois chaînes périphériques.

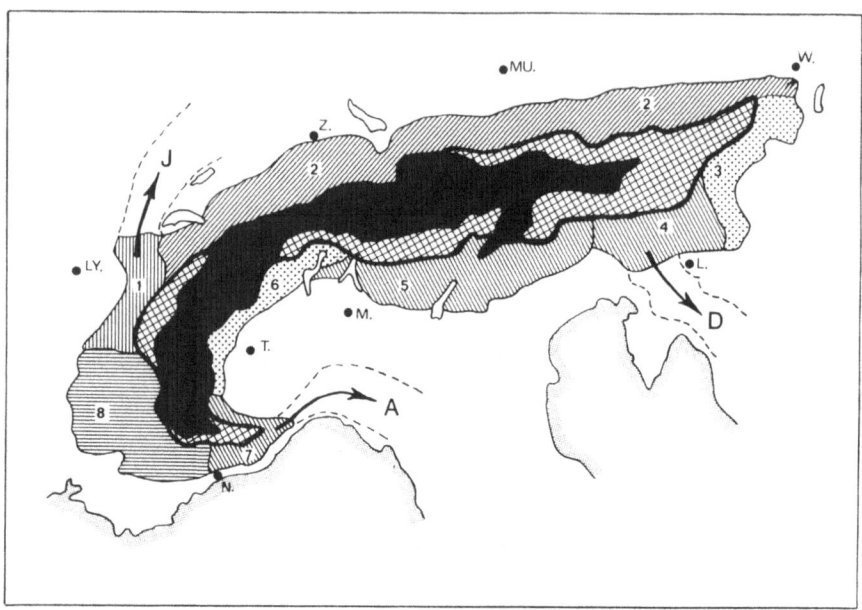

Fig. 1. Schéma des secteurs biogéographiques proposés pour la chaîne alpine (Ozenda, 1966, modifié), et de leurs relations avec les chaînes périphériques. En noir, l'axe intra-alpin proprement dit; en quadrillé, la couronne des 'Zwischenalpen' en Autriche et les écailles correspondantes dans les Alpes occidentales; en hachures, les secteurs préalpins: 1, secteur delphino-jurassien se continuant en J dans le Jura méridional; 2, secteur des Préalpes nord-occidentales; 3, secteur suprapanonnique; 4, secteur illyrique se continuant en D dans les Dinarides; 5, secteur gardésan-dolomitique; 6, secteur insubrien-piémontais; 7, secteur préligure se prolongeant en A dans l'Apennin du Nord; 8, secteur haut-provençal. Les secteurs à prédominance de roches calcaires sont figurés en hachures. Les principales villes sont indiquées par leur initiale.

1. Préalpes du Nord et Jura

On doit à Bartoli (1962) le premier travail phytosociologique précis sur un massif préalpin occidental, décrivant en Chartreuse six associations forestières de l'étage montagnard et une de l'étage subalpin. Peu après, une carte détaillée au 1/50 000 de la végétation du même massif était établie (Clerc, 1964; Ozenda et Coll., 1964; L. Richard, 1971), suivie de celle du Vercors (Faure, 1968) et d'une revue d'ensemble sur les Préalpes nord-occidentales (L. Richard, 1970). La comparaison avec des travaux jurassiens antérieurs (Luquet & Aubert, Quantin, Pottier-Alapetite) et avec ceux plus récents de J.-L. Richard, 1961, 1965, 1966, de Gehu et Coll., 1972 et surtout la comparaison du *Querceto-Buxetum* et de ses formations thermophiles, celle aussi des formations subalpines dans les deux chaînes, ont alors conduit à un parallélisme qu'illustre le Tableau I, et à la notion d'un secteur phytogéographique unique (fig. 2). Les différences entre Préalpes et Jura sont, à l'intérieur de ce secteur, relativement accessoires et tiennent d'une part à la dispo-

Tableau I – Correspondence entre les biocénoses forestières du Jura et des Préalpes occidentales. Les chiffres sont ceux que portent les biocénoses dans les mémoires de L. Richard et J.-L. Richard respectivement.

PREALPES NORD-OCCIDENTALES (L. Richard , 1966)	JURA SUISSE OCCIDENTAL (J.-L. Richard, 1961 et 1965)	LOCALISATION DANS LE JURA OCCIDENTAL
COLLINEEN HYGROPHILE		
Série de Alnus glutinosa (2)	Pruno-Fraxinetum (2)	Plaines de l'Avant-Pays
Série alluviale de Quercus robur (4)	Querco-Carpinetum p.p. (1)	
COLLINEEN SUBXEROPHILE		
Série de Quercus pubescens (5)	Coronillo-Quercetum (5)	Rebord du Premier Plateau
	Buxo-Quercetum, Querco-Lithospermetum	
COLLINEEN MESOPHILE		
Série de Carpinus betulus (9)	Querco-Carpinetum p.p. (1)	Premier gradin du Premier Plateau
Faciès à Fraxinus et Ulmus	Aceri-Fraxinetum (3)	
Faciès à Fagus (submontagnard)	Tilio-Fagetum (16)	
Série acidiphile de Quercus petraea	Lathyro-Quercetum (6)	
Série acidiphile de Fagus (submontagnard)	Melampyro-Fagetum (8), Luzulo-Fagetum (9)	
MONTAGNARD		
Série thermophile de Fagus (6)	Carici-Fagetum (11) et Seslerio-Fagetum (15)	Deuxième gradin du Premier Plateau
Faciès à Pinus silvestris (6 a)	Daphno-Pinetum (25) (= Coronillo-Pinetum)	
Série de la Hêtraie-Sapinière (10)		Rebord du Deuxième Plateau
Niveau inférieur à Fagus	Fagetum silvaticae (12)	
Niveau moyen à Fagus et Abies	Milio-Fag. (10), Abieti-Fag. (13), Equiseto-Abiet. (20)	
Niveau supérieur à Fagus et Picea	Asplenio-Piceetum p.p. (22)	Deuxième Plateau
Faciès à Acer pseudoplatanus	Aceri-Fagetum (14)	
SUBALPIN		
Série subalpine de Picea abies (13)	Sphagno-Piceetum (21)	Haute chaîne
	Asplenio-Piceetum p.p. (22), Tofieldio-Piceetum (24)	
Série de Pinus uncinata (14)	Lycopodio-Pinetum (23) (= Lycopodio-Mugetum)	

sition en bandes assez régulières (colonne 3 du Tableau I) des étages dans le Jura, disposition elle-même liée à une topographie plus simple de cette chaîne, d'autre part au moindre développement de l'étage subalpin du Jura, tant en altitude et en surface couverte qu'en richesse floristique et biocénotique (Beguin). Il faut toutefois souligner que la végétation de l'étage subalpin (à partir de 1500 m environ dans les Préalpes et de 1400 m dans le Jura; on sait que dans ce dernier les principaux reliefs sont situés précisément dans sa partie sud-occidentale) reste cependant très semblable dans les deux chaînes puisque l'on y retrouve une mosaïque formée par la Pessière subalpine sur sol profond et l'association à Pin à crochets sur lithosol, des Mégraphorbiaies à Adénostyle dans les fissures karstiques, les mêmes pelouses à *Carex sempervirens* et *Sesleria varia.*

Cette similitude entre les deux chaînes est encore renforcée par l'existence de chaînons qui, prolongeant les derniers plis du Jura vers le Sud (montagnes du Chat et de l'Epine en particulier, bordant à l'Ouest le Lac du Bourget) vont s'accoler à la bordure occidentale de la Chartreuse (fig. 2), assurant ainsi la continuité des formations collinéennes et montagnardes dans tout le secteur.

A l'Est du secteur delphino-jurassien, la nature de la végétation se modifie quelque peu, mais les étages classiques restent parfaitement reconnaissables et leurs homologies avec ceux des parties occidentales demeurent claires. Certes, Jura oriental et Préalpes orientales ne sont pas en continuité, mais les conditions de milieu naturel y sont, à une faible différence de latitude près, sensiblement identiques. L'étage montagnard prolonge presque sans modification celui du secteur

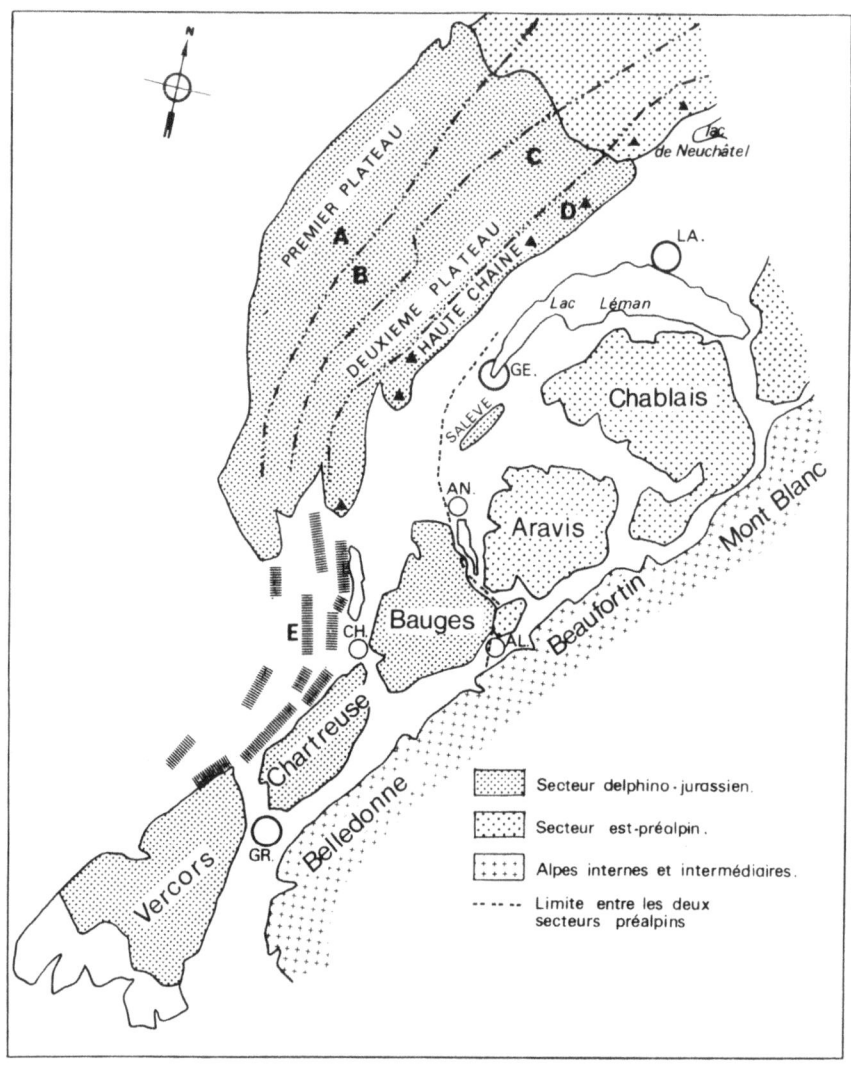

Fig. 2. Relations des Préalpes et du Jura occidental. – A à E, différentes parties du Jura occidental: A et B, premier et deuxième gradins du premier plateau; C, deuxième plateau; D, haute chaîne (les triangles indiquent les sommets dépassant 1500 m); E, chaînons liaison entre le Jura méridional, la Chartreuse et le Vercors. Le Sud du massif du Vervors appartient déjà aux Préalpes sud-occidentales. Les massifs de Belledonne, Beaufortin et Mont-Blanc sont formés de cristallin, mais Belledonne possède, sur son rebord septentrional, une importante couverture sédimentaire liasique qui n'a pas été figurée ici. Les principales villes (Lausanne, Genève, Annecy, Chambéry, Albertville, Grenoble) sont désignées par leurs deux premières lettres. Erratum: la limite entre les deux secteurs préalpins doit passer au Sud du Salève (et non au Nord).

delphino-jurassien, avec la trilogie altitudinale *Cephalanthero-Fagion, Abieti-Fagion* et *Aceri-Fagion*, mais avec un développement cependant plus grand des formations acidophiles (J.-L. Richard, 1961). Des modifications plus importantes interviennent au niveau de l'étage collinéen, où les Chênaies pubescentes sont relativement rares, leurs faciès thermophiles (et notamment la Chênaie à Buis) pratiquement absents, tandis que les Chênaies à Charme et les formations associées à Tilleul et autres feuillus (Ceinture de la forêt mixte de Chêne, de Tilleul et d'Erable de Schmid) deviennent la règle. L'étage subalpin fait pratiquement défaut, faute d'une altitude suffisante, dans le Jura suisse; mais dans les Préalpes suisses, il se poursuit avec des caractères analogues à ceux du secteur occidental et ce n'est que beaucoup plus à l'Est qu'apparaît le Pin mugo que nous retrouverons comme caractéristique de tous les secteurs décrits ci-après. Nous ne nous étendrons pas davantage sur cette comparaison entre Jura et Préalpes de l'Est, les divisions phytogéographiques de ces dernières étant encore mal établies (et la bibliographie qui les concerne, particulièrement abondante et rapidement croissante!).

Nous ajouterons cependant deux remarques: a) la limite entre le secteur delphino-jurassien et le secteur oriental peut être tracée avec une certaine précision à travers les Préalpes de Savoie et de Haute-Savoie: elle paraît, en l'état actuel des connaissances, traverser l'Est du Massif des Bauges sensiblement suivant une ligne Albertville-Annecy, puis contourner par l'Ouest le Massif du Chablais. Dans ce dernier, les formations à Buis classiquement citées n'appartiendraient pas à un véritable *Querceto-Buxetum*, mais simplement à un faciès plus sec de la Chênaie sessile. Au Nord du Lac Léman, à travers le Jura, la limite des deux secteurs nous paraît pour le moment moins claire; b) il faut rapprocher du secteur delphino-jurassien les reliefs de la Côte d'Or, formant au Nord de Dijon un petit chaînon dans lequel le Collinéen a tous les caractères du Jura occidental et qui s'élève assez pour comporter des Hêtraies submontagnardes. D'un autre côté, les Juras de Souabe et de Franconie ont certainement des rapports phytogéographiques étroits avec le Jura oriental proprement dit, (cf. la carte de Müller & Oberdorfer, 1974), mais à défaut d'expérience personnelle de ces deux régions, nous les laisserons ici de côté.

2. Préalpes du Sud et Apennin

A. Pendant longtemps, jusque vers 1950, les vues des phytogéographes sur les Alpes méridionales n'étaient pas très claires. C'est ainsi que l'on réunissait sous le nom de 'secteur insubrien' une vaste bande correspondant à toute la région des lacs de Haute Italie, du Lac d'Orta jusqu'au Lac de Garde; par ailleurs, les Alpes maritimes et ligures n'étaient pas nettement délimitées ni à droite ni à gauche, les Alpes ligures étant souvent réunies à l'Apennin; quant aux Préalpes piémontaises, vicentines, ou vénitiennes, elles n'avaient été pratiquement étudiées que sous l'angle floristique.

Une première mise au point sur les Préalpes gardésanes et dolomitiques a été

publiée en 1956 à l'occasion de la XIème Excursion Phytogéographique Internationale, bien que la délimitation des étages de végétation n'ait pas été formellement explicitée. Une description plus précise de ces étages dans les mêmes régions est donnée par Pitschmann et Coll. en 1959; leurs indications, modifiées quelque peu en fonction de mes observations personnelles, peuvent se résumer dans la succession altitudinale suivante: − un étage collinéen complexe, où s'intriquent des groupements acidophiles à *Castanea*, et calciphiles à *Ostrya* et *Quercus pubescens*, avec remontée dans la base du Collinéen calciphile de quelques transgressives méditerranéennes dont le Chêne vert et les célèbres Olivettes du Lac de Garde (étage supraméditerranéen inférieur, et non méditerranéen véritable);
− un étage montagnard classique, à Hêtraies ou Hêtraies-Sapinières;
− un Subalpin presque asylvatique, dominé par les brousses de Pin mugo.

Cet étagement est très différent de celui que l'on connaît dans les Alpes occidentales proprement dites, par exemple en Haute-Provence. Mais il est très voisin, par contre, de celui des Alpes maritimes et ligures (Ozenda, 1966 et 1969) en raison de l'importance que prennent dans ces dernières les éléments floristiques et phytosociologiques orientaux dont beaucoup s'y trouvent à leur limite occidentale (Ostryaies, Pinèdes à *Erica carnea*, Pin mugo); d'où le rapprochement (voir plus haut, fig. 1) entre le secteur gardesan dolomitique et un secteur préligure qui en est la réplique dans des conditions naturelles assez semblables, 200 km au Sud-Ouest. Il faut noter que le secteur préligure considéré ici comprend l'ensemble des Préalpes maritimes orientales et des Alpes ligures, mais non la partie interne (Mercantour et annexes) des Alpes maritimes.

Dans une analyse phytosociologique détaillée de l'étage collinéen des Préalpes lombardes, c'est-à-dire du secteur insubrien *sensu lato* des anciens auteurs, Oberdorfer (1964) a distingué nettement: a) un secteur insubrien proprement dit (réuni dans la fig. 1 aux Préalpes piémontaises), à l'Ouest du Lac de Lugano et jusqu'-aux environs de Biella, dans lequel le climax est formé d'associations dépendant du *Carpinion* et du *Quercion roboris*, affines de la végétation médio-européenne du versant Nord des Alpes, avec introduction massive de plantes cultivées de type laurifolié dans les parties riveraines des lacs; b) un secteur gardésan, qu'occupe au contraire, sur sol calcaire prédominant, un complexe d'affinités subméditerranéennes, tant dans sa végétation naturelle que dans les cultures.

B. Etudiant par la suite l'Apennin ligure qui présente une intrication compliquée de terrains calcaires et siliceux, Oberdorfer a également montré (1967) que l'on y retrouve dans les étages collinéen et montagnard la mosaïque des deux complexes précédents, avec leurs associations reconnaissables, l'exposition jouant en outre un rôle dans leur répartition. Barbero & Bono (1970, 1973) ont abouti aux mêmes conclusions, qu'ils ont pu étendre d'ailleurs à l'étage subalpin, dans les Alpes apuanes. D'après les travaux de ces trois auteurs, on peut dresser le tableau suivant (Tableau II), qui ne comprend pas l'étage subalpin dont il sera question plus loin à propos de l'Apennin central.

D'autres traits alpins se retrouvent dans la végétation de l'Apennin septentrio-

Tableau II

ETAGES et altitudes approximatives	TYPE MEDITERRANEEN Roches carbonatées, ou exposition Sud	TYPE MEDIOEUROPEEN Roches siliceuses, ou exposition Nord
1600 MONTAGNARD 900	Hêtraie thermophile d'altitude ⟍ Aceri-Fagetum Hêtraie neutrophile (Cardamino heptaphyllae-Fagetum) Hêtraie thermophile (Carici-Fagetum)	Hêtraie acidiphile (Luzulo pedemontanae-Fagetum)
SUPRAMEDIT. 200	Chênaie à Quercus cerris - - - - - - - - - - Orno-Ostryetum - - - - - - - - - - Orno-Quercetum pubescentis	Chênaie acidiphile (Physospermo-Quercetum), Salvio-Fraxinetum et divers groupements du Carpinion
MESOMEDIT.	Orno-Quercetum ilicis	Quercetum "mediterraneo-montanum"

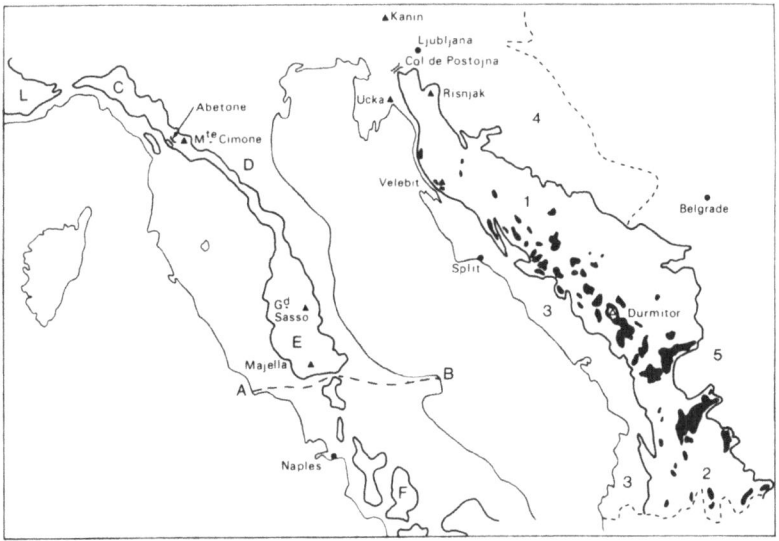

Fig. 3. Position des principaux massifs de l'Apennin et des Dinarides cités dans le texte. Le contour en trait épais représente dans les deux chaînes la limite approximative du complexe montagnard-subalpin, c'est-à-dire celle des Hêtraies.

L, Alpes ligures, A-B, limite biogéographique entre l'Italie centrale et l'Italie du Sud, correspondant notamment à la limite septentrionale de l'*Oleo-Ceratonion* sur la côte et à celle des Hêtraies méditerranéennes en montagne. 1 et 2, étage montagnard des Dinarides septentrionales et méridionales; l'étage subalpin correspond aux taches noires. 3, Collinéen de type méditerranéen à *Ostrya* et *Carpinus orientalis* (*Orno-Ostryetum adriaticum*); 4 et 5, Collinéen de type médio-européen, formé de groupements du *Carpinion* en 4 et de Chênaies subcontinentales en 5. (Pour la partie dinarique, d'après Horvat, Glavac & Ellenberg, simplifié).

25

nal. Rappelons l'existence de la très belle Sapinière du massif de l'Abetone (fig. 3), celle de peuplements relictuels mais assez importants de *Picea* et de *Vaccinium myrtillus* dans le même massif, de *Pinus mughus* et de *Rhododendron ferrugineum* au Monte Cimone, de *Alnus incana* çà et là (Bertolani-Marchetti).

C. La végétation de l'Apennin central est relativement bien connue par les travaux de Lüdi, Hofmann, Marchesoni, Tomaselli, Pedrotti et Coll., entre autres. Avec l'éloignement des Alpes et la latitude décroissante, l'influence méditerranéenne est prédominante dans les étages inférieurs. L'étage collinéen devient ici un Supraméditerranéen typique, avec des formations rupicoles à Chêne vert ('Leccete impoverite') et des influences orientales (*Quercus cerris, Carpinus orientalis*). Dans l'étage montagnard cependant, les Hêtraies conservent, malgré une dégradation accentuée de leurs niveaux inférieurs, un caractère septentrional dont témoigne leur composition floristique (*Asperulo-Fagion, Dentario-Fagetum* bien reconnaissables) et la présence du Sapin. Mais c'est dans les étages supérieurs, favorisés par l'altitude plus élevée de cette partie de la chaîne qui permet même l'existence d'un étage alpin, que les caractères septentrionaux se retrouvent avec netteté. L'étage subalpin comporte à vrai dire, comme dans les Alpes apuanes (et aussi, sous une forme différente, en Corse) une intrication de deux complexes:

a) des groupements de type médio-européen affines de ceux des Alpes méridionales: groupements d'éboulis à *Festuca dimorpha*, combes à neige, pelouses à *Festuca violacea* et *Trifolium thalii*, à *Juncus trifidus*, à *Antennaria dioica*, pâturages siliceux à *Festuca spadicea*; les importantes étendues de brousse à Pin mugo du Massif de la Majella sont également caractéristiques;

b) des groupements de type oroméditerranéen, en partie affines de ceux des Balkans (Lakusic, 1969), comme la pelouse à *Sesleria tenuifolia* var. *apennina* très riche en espèces endémiques.

L'étage alpin est bien connu aussi pour ses endémiques, dont beaucoup sont des vicariantes d'espèces alpines (*Leontopodium nivale* est la plus célèbre) et pour ses arctico-alpines (*Salix retusa*, diverses Saxifrages); mais ici également l'origine géographique double est perceptible, attestée par la présence d'un endémisme de type oroméditerranéen, en partie apennino-balkanique, et portant notamment sur les Campanulacées (*Hedreanthus*).

D. Avec l'Apennin méridional, la coupure est brutale, tant sur le plan floristique que biocénotique. Moggi a bien mis en évidence la discontinuité floristique par une statistique portant sur quatre familles à prédominance d'Orophytes. La rupture la plus nette se produit certainement au niveau des Hêtraies, où les associations de type médio-européen sont remplacées par des groupements méridionaux qui en sont assez différents pour que l'on ait proposé de les ranger, sous le nom de *Geranio-Fagion*, dans une alliance distincte. Nous sommes là dans les montagnes méditerranéennes proprement dites, dans lesquelles l'étagement ne répond plus du tout au modèle alpin.

3. Préalpes sud-orientales et Dinarides

Le contact entre Préalpes et Dinarides se situe au centre de la Slovénie, où les deux chaînes encadrent le bassin de Ljubljana et ne sont séparées que par le Col de Postojna (Adelsberg) à 1500 m.

A. Par chance, les Préalpes slovènes sont relativement bien connues, grâce à de nombreux travaux dont ceux de M. Wraber, Puncer & Zupancic (1970). La région de Bovec, en particulier, permet de voir de nettes séquences d'étagement altitudinal dont la coupe du versant Sud du Kanin est un des meilleurs exemples. On y trouve de bas en haut (observations personnelles): — une formation planitiaire-collinéenne, complexe de groupements mal caractérisés, appartenant peut-être au *Carpinion* et au *Quercion pubescentis*, qui occupe les alluvions des fonds de vallées et les sols colluviaux de bas de pentes; — un Collinéen supérieur, sur les premiers escarpements calcaires, fait d'un *Orno-Ostryetum* passant progressivement, avec l'altitude, à un *Ostryo-Fagetum* submontagnard; — une Hêtraie-Sapinière (*Anemone-Trifolio-Fagetum abietetosum*), entre 900 et 1550 m environ; — un Subalpin à Pin mugo, qui n'atteint pas son amplitude altitudinale, à cause de l'extension exceptionnelle des falaises rocheuses qui le limitent à sa partie supérieure, falaises dont les vires et les replats sont colonisés par un *Seslerieto-Semperviretum*.

Au total, un type d'étagement qui, si ce n'était la présence de l'Ostrya dans une partie de l'étage collinéen, pourrait être celui des Préalpes nord-orientales: nous reviendrons sur ce point. Il est assez général dans toutes les Alpes slovènes, exception faite cependant d'importantes étendues d'Epicéa et de Pin sylvestre, et des formations silicicoles particulières au Massif du Pohor.

B. Moins puissantes, tronquées au sommet de l'étage montagnard, les Dinarides slovènes sont néanmoins la réplique parfaite des Préalpes qui leur font face. Des divergences de nomenclature phytosociologique, reflet inévitable des différences régionales dans un pays à la flore aussi riche, ne parviennent pas à masquer la quasi-identité des groupements de part et d'autre (Sugar, 1970; Trijnastic, 1970): mêmes Ostryaies, mêmes Hêtraies à *Hacquettia*, à Sapin (*Abieti-Fagetum dinaricum*) ou à Sycomore (Zupancic, 1969), mêmes dolines à inversion d'étages tapissées de *Piceetum* à flore froide dans leur fond.

C. La même structure se poursuit à travers les montagnes croates, jusqu'au Montenegro, où la coupe du Durmitor, massif culminant de la chaîne des Dinarides, est classique depuis les travaux de Horvat (fig. 4). La similitude avec les Préalpes reste profonde: ainsi le tableau du dynamisme de la végétation à l'étage du Pin mugo, que donne Horvat (1962, p. 52) montre les groupements typiques du Subalpin calcicole: *Firmetum, Seslerieto-Semperviretum*. La relative régularité structurale de la chaîne entraîne la disposition des complexes de végétation, correspondant aux étages, en bandes parallèles à la côte (fig. 3): un axe de Montagnard à Hêtraie-Sapinière, de plus de 150 km de largeur par endroits, que ponctuent les taches de Subalpin à Mugo et *Sesleria* des parties les plus élevées; de part et d'autre, deux étages collinéens sensiblement différents, un *Ostryo-Carpinetum orientalis* sur le versant adriatique, un *Carpinion* de type médio-européen du côté

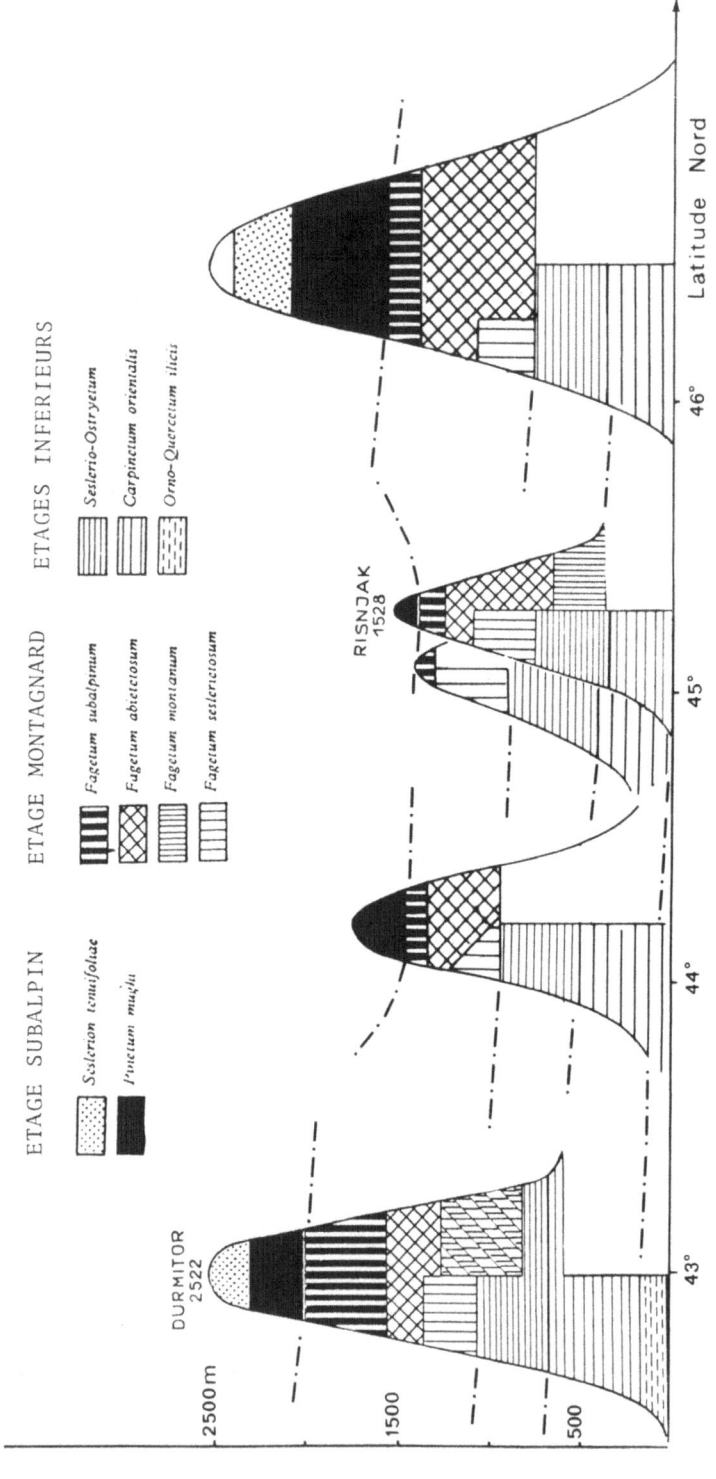

Les cinq massifs représentés sont de gauch à droite, c'est-à-dire du Sud vers le Nord, le Durmitor, le Velebit, l'Ucka-Gora, le Risnjak et les Alpes slovènes.

opposé (alors que ces deux types de Collinéen sont intriqués dans les Préalpes slovènes).

D. Ce n'est qu'aux approches de l'Albanie que les choses changent: le *Carpinion* cède la place aux Chênaies subcontinentales caractéristiques de l'Europe du Sud-Est, le Hêtre des Balkans (*Fagus moesiaca*) se substitue progressivement au Hêtre commun, le Pin mugo est remplacé par *Pinus heldreichii* et *P. peuce*. Il faudra pourtant atteindre le territoire grec et la chaîne du Pinde pour voir le schéma fondamental des étages, tel que nous le suivions depuis les Préalpes, s'altérer sensiblement, et aller plus loin encore, au niveau de la Phocide, pour atteindre les véritables montagnes eu-méditerranéennes, symétriques de l'Apennin du Sud.

Ainsi, le 'modèle alpin' s'avance dans les Balkans plus au Sud que dans l'Apennin, probablement en raison de la morphologie plus massive des montagnes de la péninsule orientale, car, avec des sommets pourtant un peu moins élevés, les Dinarides forment un bloc beaucoup plus large et moins morcelé que la Dorsale italienne.

4. A la recherche d'une représentation unitaire

Comme nous venons de le voir, la végétation de chacune des trois chaînes périphériques est en quelque sorte 'ancrée' dans un secteur déterminé des Préalpes. Mais nous pouvons aller plus loin en reprenant les analogies que les divers secteurs préalpins présentent entre eux, et qui ont été évoquées notamment à propos de la similitude entre la grille de végétation des Préalpes slovènes et celle des Préalpes de Bavière ou d'Autriche. La Fig. 5 illustre ces homologies entre tous les massifs

---

Fig. 4. Etagement comparé de la végétation dans les Dinarides et les Préalpes sud-orientales. Pour le Durmitor et le Risnjak, on a conservé sans modification les coupes données par Horvat (1962), ainsi que ses notations; la coupe du Velebit a été établie d'après Trijnastic, celle de l'Ucka-Gora d'après Zugar, celle des Alpes slovènes d'après différents auteurs et des observations personnelles. Pour ces trois derniers massifs, on a employé les mêmes représentations que Horvat.

Les schémas sont présentés du Sud (à gauche) au Nord (à droite) en raison de l'orientation donnée par Horvat à ses figures. Les limites d'étages s'abaissent naturellement du Sud vers le Nord, mais les documents disponibles ne permettent pas d'évaluer exactement le gradient correspondant (rappelons qu'il est en général de 100 m d'altitude environ par degré de latitude). Les groupements des Alpes slovènes et du Nord des Dinarides qui ne pénètrent pas plus au Sud n'ont pas été figurés (*Carpinion, Aceri-Fagetum, Laricetum*).

Remarques particulières: 1°/ L'étage mésoméditerranéen à *Orno-Quercetum* se lamine en bordure de la mer vers 45°. 2°/ Le *Fagetum subalpinum* des auteurs yougoslaves représente pour nous le sommet de l'étage montagnard. 3°/ Dans les Alpes slovènes: a) l'étage collinéen en versant Sud comprend à sa base des groupements subxérophiles dont certains sont voisins du *Quercetum pubescentis; Carpinus orientalis* en est absent; b) en versant Nord ou sur sol profond, le Collinéen est formé de groupements du *Carpinion;* c) dans le Montagnard moyen en exposition Sud, le Sapin est souvent remplacé par l'Epicéa; d) le Montagnard supérieur comprend une proportion notable de Mélèze.

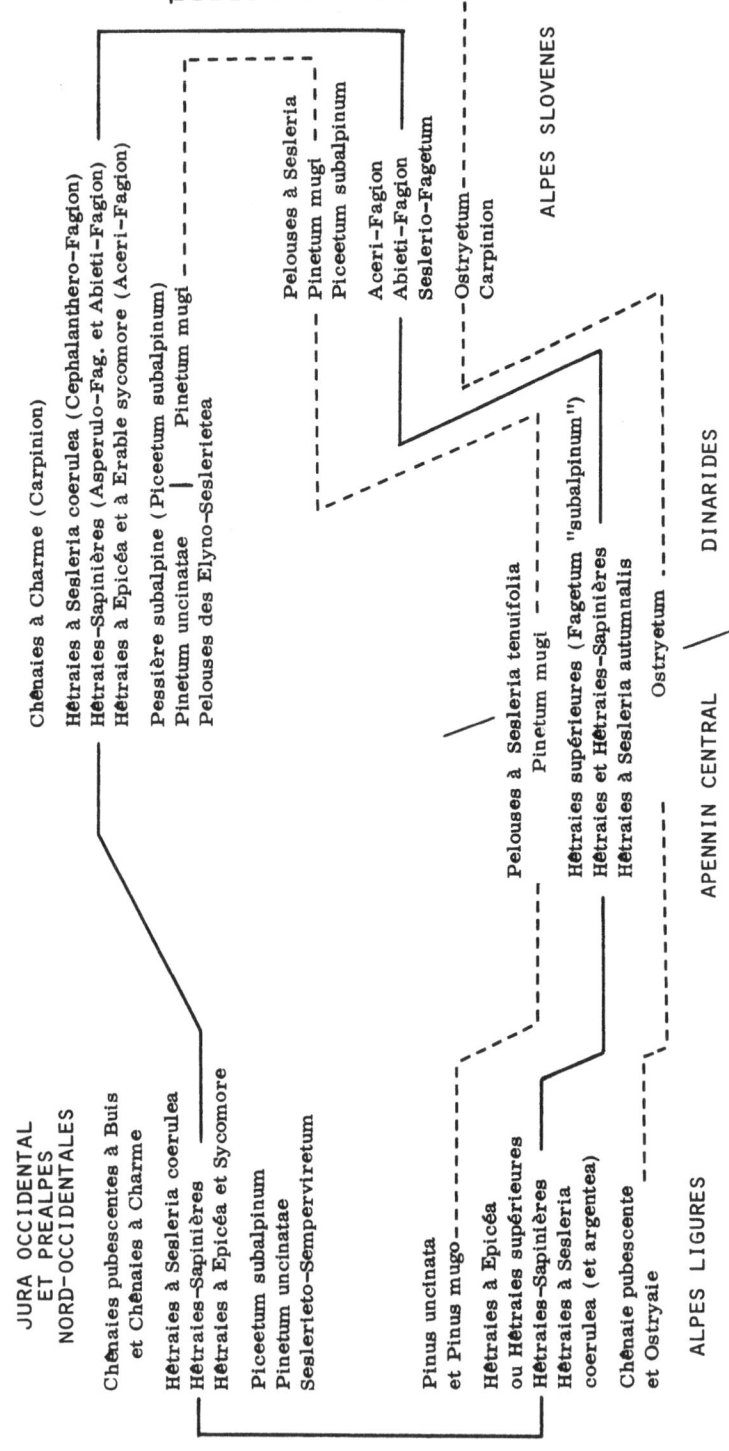

30

considérés ici (on pourrait y ajouter, mais ce point n'a pas été traité, l'évidente relation entre les Préalpes orientales et certaines parties des Carpates nord-occidentales).

Le bloc le plus homogène est formé par l'ensemble constitué de la partie sud de l'Arc alpin allant de la Basse Autriche aux Alpes maritimes, de l'Apennin septentrional et central, et des Dinarides. La grille de l'étagement de végétation y est remarquablement uniforme: les Hêtraies montagnardes, structurées en trois sous-étages constants, sont encadrées par le Subalpin à Pin mugo et par un Collinéen à Ostrya qui pénètre localement jusqu'en Styrie.

Enfin, entre Apennin central et Dinarides, la conformité de l'étagement est si étroite que nous pouvons réunir ces deux chaînes, dans la Fig. 5, en une grille unique. Le détail de la comparaison phytosociologique renforcerait sur plus d'un point ces analogies (Lakusic, 1969). Il serait intéressant d'examiner si cette uniformité des chaînes calcaires s'étend aussi à la comparaison de leurs régions siliceuses, qui sont plus limitées: la similitude de l'aire de nombreuses espèces, et en particulier d'espèces aborescentes importantes à préférences silicicoles, comme *Quercus cerris* et *Castanea vesca*, permet de le penser (fig. 6). Ainsi se trouverait davantage encore confirmée notre hypothèse d'un grand système montagneux péri-adriatique se prolongeant vers le Nord-Ouest jusque dans les Alpes maritimes et vers le Nord-Est au moins jusqu'en Carinthie.

Les considérations qui précédent sont indépendantes de toute hypothèse sur l'origine et sur l'historique de ce complexe biogéographique préalpin-périalpin. En effet, dire que la végétation alpine 'se prolonge' dans les chaînes calcaires péri-alpines ne préjuge en rien du sens dans lequel a pu sa faire la mise en place des ensembles concernés. Il est même plus vraisemblable que cette mise en place s'est faite *à partir* des chaînes périphériques qui auraient servi de refuge pendant les glaciations, donc en direction des Préalpes et non pas à partir d'elles. La richesse beaucoup plus grande de la flore balkanique est précisément un témoin de ce rôle de territoire de refuge. L'appauvrissement relatif, par rapport aux Préalpes, de l'Apennin central serait un phénomène secondaire dû à l'isolement et aux faibles dimensions de ce massif; Pignatti pense d'ailleurs que les Hêtraies de l'Apennin du Sud ne dérivent pas de celles de l'Apennin du Nord et des Préalpes par appauvrissement, mais qu'elles en ont été au contraire le point de départ lors de la recolonisation post-glaciaire.

Fig. 5. Comparaison entre les étagements de végétation de l'ensemble des massifs considérés. Les traits épais qui réunissent entre elles les Hêtraies-Sapinières de l'étage montagnard moyen schématisent l'unité de cet étage dans tout l'ensemble. Les tirets représentent la distribution de *Ostrya carpinifolia* et des *Sesleria* du groupe *argentea* et *autumnalis* dans l'étage collinéen, et du Pin mugo dans l'étage subalpin; ils matérialisent la liaison entre l'ensemble des massifs du système péri-adriatique proposé. Pour chaque massif, les étages et sous-étages de végétation sont énumérés en allant du Collinéen, orienté vers la périphérie de la figure, au Subalpin orienté vers le centre de la figure.

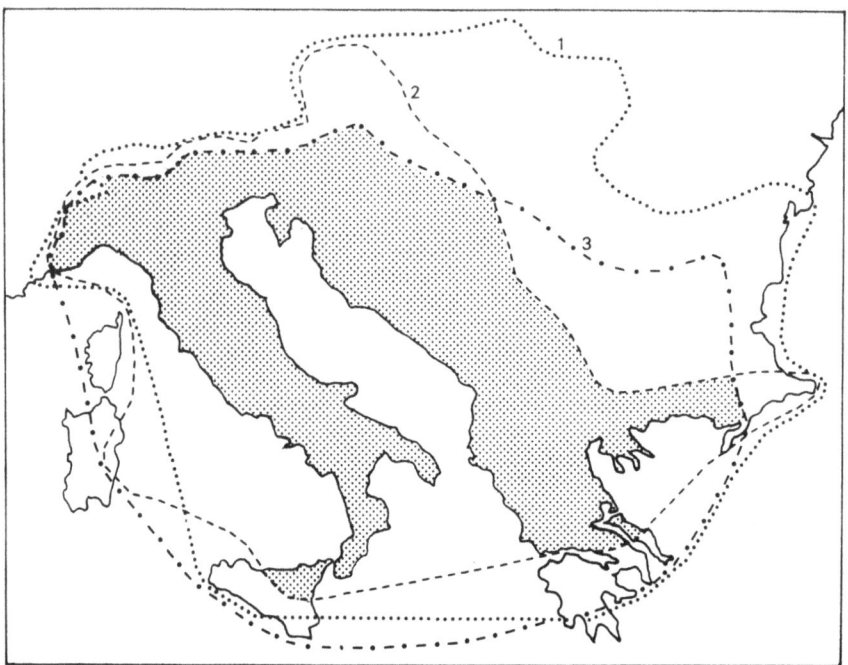

Fig. 6. Limites de *Quercus cerris* (1), *Castanea vesca*, aire de l'espèce à l'état spontané (2), *Ostrya carpinifolia* (3), en pointillé, les parties communes aux trois aires. (d'ap. Jalas et Suominen) Beaucoup d'autres espèces ont une répartition du même type, correspondant à ce que nous nommons dans le texte 'système péri-adriatique.'

Summary

The phytogeographical relationships between the Alpes and the peripheric limestone ranges (Jura, Apennin, Dinarides). — From comparative interpretation of the altitudinal patterns of the vegetation climax in the different sectors of the Prealps and in the three big peripheric ranges (fig. I), it is shown that the western parts of the northern Prealps and of the Jura belong evidently to the same phytogeographical sector (table I and fig. 2), while strong affinities appear between the Ligurian Alps and the northern and central Apennin and on the other hand between Julian Alps and Dinarids (fig. 3). Moreover, the relations between all parts of the whole prealpine and perialpine limestone system lead to a single model whose montane level exhibits a remarkable uniformity, but which split in two undertypes after consideration of the framing levels and especially of the respective distribution Carpinion/Ostryon and Pinus uncinata/Pinus mugo (fig. 5). Finally, the close parallelism between Apennin and Dinarids and other chorologic remarks (fig. 6) allow to introduce the concept of a peri-adriatic mountain system which seems even include the major part of the southern Prealps and may be the geographical origin of the whole vegetation complex discussed here.

Bibliographie

Barbero, M. & Bono, G. 1970. La végétation sylvatique thermophile de l'étage collinéen des Alpes apuanes et de l'Apennin ligure. – Lavori della Soc. It. di Biogeogr., N.S.I., 147-182.

Barbero, M., Bono, G. & Ferrarini, E. 1970. Le Alpi apuane: i loro rapporti con le Alpi marittime et liguri, l'Appennino settentrionale, le Alpi orientali e Dinariche. – Arch. Bot. et Biog. Ital., XLVI, 135-152.

Barbero, M., Bono, G. & Ozenda, P. 1970. Sur les groupements végétaux en limite d'aire dans les Alpes maritimes et ligures. – Bull. Soc. Bot. Fr., 117, 593-608.

Barbero, M. & Bono, G. 1973. La végétation orophile des Alpes apuanes. – Vegetatio, 27: 1-48.

Bartoli, Ch. 1962. Associations forestières du Massif de la Grande Chartreuse. – Ann. Ecole Nat. E. et F., XIX, 327-383.

Beguin, Cl. 1970. Contribution à l'étude phytosociologique et écologique du Haut-Jura. – Thèse, Neuchâtel, 189 p., 1 carte coul.

Bertolani-Marchetti, D. 1962. L'ambiente botanico degli itinerari al Monte Cimone. – Guido dell'Alto Appennino, 3-15.

Clerc, J. 1964. Feuille de Grenoble (XXXII-34). – Doc. Carte Vég. Alpes, II, 37-68, 13 fig., 1 carte coul. 1/50 000.

Dobremez, J.-F. 1972. Mise au point d'une méthode d'étude cartographique des montagnes tropicales. Le Népal, écologie et phytogéographie. – Univ. Grenoble, Thèse, 373 p., 180 fig., 25 tabl., 23 phot., cartes coul.

Faure, Ch. 1968. Feuille de Vif (XXXII-35). – Doc. Carte Vég. Alpes, VI, 7-69, carte coul., 8 tabl., 8 phot.

Gehu, J.-M., Richard, J.-L. & Tüxen, R. 1972. Compte-rendu de l'excustion de l'Association Internationale de Phytosociologie dans le Jura en Juin 1967. – Doc. Phytosoc., I, 1-44, et II, 1-50.

Hofmann, A. 1974. Dalle Madonie alle Alpi Giulie attraverso le faggete italiane. – Notiz. della Soc. Ital. di Fitosoc., 9: 3-14.

Horvat, I. 1962. La végétation des montagnes de la Croatie occidentale. – Prirodoslovna Istrazivanja, 30: 1-179.

Horvat, I., Glavac, V. & Ellenberg, H. 1974. Vegetation Südosteuropas. – Stuttgart, Fischer, 768 p.

Jalas, H. & Suominen, J. 1976. Atlas Florae Europae. – Helsinki, vol. 3, 128 p.

Lakusic, R. 1969. Vergleich zwischen den Elyno-Seslerieta Br.-Bl. der Apenninen und Dinariden. – Mitteil. d. Ostalpin-dinar. Pflanzensoz. Arbeitsgem., 9: 133-143.

Lüdi, W. 1944. Die Gliederung der Vegetation auf der Apenninenhalbinsel, insbesondere der montanen und alpinen Höhenstufen. – In Rikli, Pflanzenkleid der Mittelmeerländer, 573-596.

Luquet, A. & Aubert, S. 1930. Etudes phytogéographiques sur la chaîne jurassienne. Recherches sur les associations végétales du Mont Tendre. – Allier, Grenoble, 50 p.

Müller, Th. & Oberdorfer, E. 1974. Die potentielle natürliche Vegetation von Baden-Württemberg. – Beihefte der Landesst. f. Natursch. und Landschaftspf. Baden-Württ., 6, 3-46, carte coul. 1/900 000.

Oberdorfer, E. 1964. Der insubrische Vegetationskomplex, seine Struktur und Abgrenzung gegen die submediterrane Vegetation in Oberitalien und in der Südschweiz. – Ber. Naturk. Forsch. SW Deutschl., XXII, 141-187.

Oberdorfer, E. 1967. Beitrag zur Kenntnis der Vegetation des Nordapennins. – Beitr. Naturk. Forsch SW Deutschl., 26: 83-139.

Ozenda, P. 1966. Perspectives nouvelles pour l'étude phytogéographique des Alpes du Sud. – Doc. Carte Vég. Alpes, IV, 10198, 4 cartes coul.

Ozenda, P. 1969. Sur la valeur biogéographique des groupements à Pin mugo dans les Alpes occidentales. – C.R. Soc. Biogéogr.

Ozenda, P. 1975. Sur les étages de végétation dans les montagnes du Bassin Méditerranéen. – Doc. Cartogr. Ecol., XVI, 1-32.

Ozenda, P. 1976. Les grandes lignes de la végétation du Caucase, vues par un biogéographe alpin. – Congr. Intern. Féogr. Moscou, 12: 143-145.

Ozenda, P. 1976. Les affinités biogéographiques occidentales de la chaîne himalayenne. – Col. Intern. C.N.R.S., n° 268: 69-80.

Ozenda, P. 1978. Carte de la végétation de l'Europe 1/3 000 000. – Conseil de l'Europe, Strasbourg (sous presse).

Ozenda, P., Repiton, J., Richard, L. & Tonnel, A. 1964. Feuille de Domène (XXXIII-34). – Doc. Carte Vég. Alpes, II, 69-118, 10 fig., 11 tabl., 1 carte coul. 1/50 000.

Pedrotti, F. 1969. Einführung in die Vegetation des Zentralapennins. – *Mitteil. d. Ostalpin-dinarischen Pflanzensoz. Arbeitsgem.*, 9: 21-57.

Pignatti, S. 1974. in Notiz. della Soc. It. di Fitosociol., 9, p. 12.

Pitschmann, H., Reisigl, H. & Schiechtl, H. 1959. Bilder-Flora der Südalpen. – Fischer, Stuttgart, 278 p.

Pottier-Alapetite, G. 1943. Recherches phytosociologiques et historiques sur la végétation du Jura central et sur les origines de la flore jurassienne. – Tunis, 333 p. (Stat. Int. Géobot. Médit. et Alp. Montpellier, communication n° 80).

Puncer, I. & Zupancic, M. 1970. Vergleich der Vegetationsgrenzen bzw. der Vegetationsprofile in verschiedenen Gebirgssystemen auf Karbonat- und Silikatunterlage in Slovenien. – *Mitteil. d. Ostalpin-dinarichen Gesells. f. Vegetationsk.*, 11, 187-196.

Quantin, A. 1935. L'évolution de la végétation de la Chênaie dans le Jura méridional. Univ. Paris, Thèse, 377 p.

Richard, J.-L. 1961. Les forêts acideophiles du Jura. Comm. phytogéogr. Soc. Helv. Sc. Nat., 38), 164 p., 38 fig., 10 tabl.

Richard, J.-L. 1965. Extraits de la Carte phytosociologique des forêts du canton de Neuchâtel. – (Comm. phytogéogr. Soc. Helv. Sc. Nat., 47), 48 p., 1 carte.

Richard, J.-L. 1966. Les forêts naturelles d'Epicéas et Pins de montagne du Jura. – *Bull. Soc. neuchât. Sc. Nat.*, 99: 101-112.

Richard, L. 1970. Les séries de végétation dans la partie externe des Alpes nord-occidentales. – Veröffentl. d. Geobot. Inst. ETH Zurich, H 43: 65-103, 5 fig., tabl.

Richard, L. 1971. Feuille de Montmélian (XXXIII-33). – Doc. Carte Vég. Alpes, IX, 9-78, 15 fig., 22 tabl., 4 phot., 1 carte coul. 1/50 000.

Schmid, E. 1939 à 1950. Carte de la Végétation de la Suisse au 1/200 000. – Berne, Kümmerly et Frey, 4 feuilles.

Stefanovic, V. & Fabijanic, B. 1969. Zerreiche und Waldgesellschaften mit Zerreiche des dinarischen und apennischen Gebietes. – *Mitteil. d. Ostalpin-dinarischen Pflanzensoz. Arbeitsgem.*, 9: 287-299.

Sugar, I. 1970. Das Vegetationsprofil des Ucka Gebirges. – *Mitteil d. Ostalpin-dinarischen Gesells. f. Vegetationsk.*, 11: 213-218.

Tomaselli, R. 1970. Carta della Vegetazione naturale potenziale d'Italia. – Rome, Minist. Agric. et For., 1 carte coul. au 1/1 000 000 et notice 64 p.

Trinajstic, I. 1970. Höhengürtel der Vegetation und die Vegetationsprofile im Velebit-Gebirge. – *Mitteil. d. Ostalpin-dinarischen Gesells. f. Vegetationsk.*, 11: 219-224.

Wraber, M. 1970. Zur Topographie, Okologie und Soziologie der slowenischen Urwälder. – Symp. sur les forêts vierges sudeuropéennes. – Ac. de Bosnie-Herzeg., Sarajevo, XV, livre 4, 91-102.

Wraber, M. 1970. Das submediterrane-illyrische Element in der mitteleuropäische Laubwaldvegetation Sloweniens. – *Feddes Repert.*, 81: 279-287.

Zupancic, M. 1969. Vergleich der Bergahorn-Buchgesellschaften (Aceri-Fagetum) im alpinen und dinarischen Raume. – *Mitt. ostalp. -din. pflanzensoz. Arbeitsgem.*, 9: 119-131.

Adress of the Author:
Prof. Dr. P. Ozenda
Université Scientifique et Médicale de Grenoble
Botanique et Biologie Végétale, Domaine Universitaire
38 Saint-Martin-D'Hères, France

LES VEGETATIONS A ARTHROCNEMUM PERENNE DES CÔTES ATLANTIQUES EUROPEENNES

Jean-Marie & Jeannette Géhu

Introduction

La Chénopodiacée vivace sous frutescente halophile, *Arthrocnemum perenne*, possède sur les côtes européennes une distribution méditerranéenne-atlantique. Espèce des Sansouires méditerranéennes, elle est également présenté sur les vases salées de la façade atlantique de l'Europe, du Sud de l'Angleterre jusqu'au Maroc. En France atlantique, elle existe du Cotentin jusqu'à la frontière d'Espagne.

Elle forme, à la limite de la Haute Slikke et du Schorre, une ceinture facultative étroite, et parfois dissociée, derrière les *Salicornietum* annuels variés (*dolichostachyae, fragilis, obscurae*) et les *Spartinetum* (*maritimae* et *townsendii*) mais devant la zone de l'*Halimionetum portulacoidis*. Elle tend aussi à pénétrer dans l'*Halimiono- Puccinellietum maritimae* qui peut exister également dès ce niveau.

Nous avons précédemment montré (Géhu 1973, 1975, Géhu & Géhu-Franck 1977, 1978) que les végétations à Salicorne vivace et à Obione appartenaient à la classe des *Arthrocnemetea fruticosae* plutôt qu'à celle des *Asteretea tripolium*.

Cette note, précisant les caractères phytosociologiques d'*Arthrocnemum perenne* s'inserre dans une série d'articles sur les irradiations atlantiques de la classe méditerranéenne des *Arthrocnemetea fruticosae*, et dans le cadre d'une étude générale des côtes atlantiques françaises. Nous sommes heureux de la présenter dans le livre jubilaire de Professeur J. Schmithüsen et de la lui dédier très cordialement.

L'association atlantique à *Arthrocnemum perenne*

Nom: *Puccinellio maritimae – Arthrocnemetum perennis* (Fontes 1945) J.M. Géhu 1975.

Le tableau synthétique n° 1 précise la combinaison floristique générale de l'association, sur la base de 181 relevés.

Tableau n° 1

	typicum	spartinetosum maritimae	spartinetosum townsendii	arthrocnemetosum fruticosi
Nombre de relevés:	105	29	35	12
Chiffre spécifique:	3,3	4,3	3,9	4
Car. et Diff. d'assoc.:				
Arthrocnemum perenne	V	V	V	V
Puccinellia maritima	II	III	III	III
Bostrychia scorpioides	II	III	+	+
Diff. de sous-assoc.:				
Spartina maritima		V		
Spartina towsendii			V	
Arthrocnemum fruticosum				V
Esp. des unités supérieures:				
Halimione portulacoides	III	II	II	IV
Compagnes:				
Aster tripolium	II	II	II	II
Suaeda maritima	II	II	I	I
Salicornia stricta	II	+	I	I
Spergularia media	I	I	+	
Limonium vulgare.	I	I	+	
Triglochin maritimum	+		+	

Synstructure: C'est une association pionnière pauci-spécifique, dans laquelle *Arthrocnemum perenne* domine toujours. Le rôle d'*Halimione portulacoides*, reste très discret, sauf dans quelques variantes bionomique ou dynamique. La strate épiphyte algale est très variable suivant les sites.

Synécologie et syndynamique: C'est l'association des *Arthrocnemetea fruticosi* présente dans les plus bas niveaux. Elle est en contact inférieur fréquent avec les spartinaies, associations pionnières vivaces des vases salées occidentales (*Spartinetum towsendii* à l'Ouest-Nord-Ouest, *Spartinetum maritimae* à l'Ouest Sud-Ouest) ou avec divers groupements de Salicornes annuelles de la Haute Slikke (*Salicornietum dolichostachyae, S. fragilis, S. obscurae*). Le contact supérieur se fait avec les associations de la zone à *Halimione portulacoides* (du Nord au Sud: *Bostrychio – Halimionetum, Puccinellio maritimae-Arthrocnemetum fruticosi* et *Cistancho-Arthrocnemetum fruticosi*).

Des effets de contact latéral et de mosaïque peuvent avoir lieu avec des phases primaires de l'*Halimiono-Puccinellietum maritimae*.

L'association parait favorisée, au moins vers ses limites nord, par l'existence de micro falaises entre Slikke et Schorre, par le mélange de pierres avec les tangues et par un certain mouvement marin (clapotis et courrant léger).

Synchorologie: L'association atlantique à *Arthrocnemum perenne* existe du Nord du Maroc et du Sud ouest ibérique jusqu'aux côtes méridionales de l'Angleterre (Géhu & Delzenne 1975) en passant par l'Ouest français (Corillion 1953, Géhu 1975).

Synsystématique: L'association atlantique diffère fondamentalement de sa vicariante méditerranéenne décrite en 1928 puis en 1933 par J. Braun-Blanquet, ainsi que l'a bien observé dès 1945 Fontes dans son étude des 'Salgados' de Sacavem (en bordure du Tage), par la présence de plantes du *Puccinellion maritimae*, en particulier *Puccinellia maritima*, tandis que manquent les transgressives méditerranéennes comme *Aeluropus littoralis* et *Puccinellia festucaeformis* (Géhu 1975).

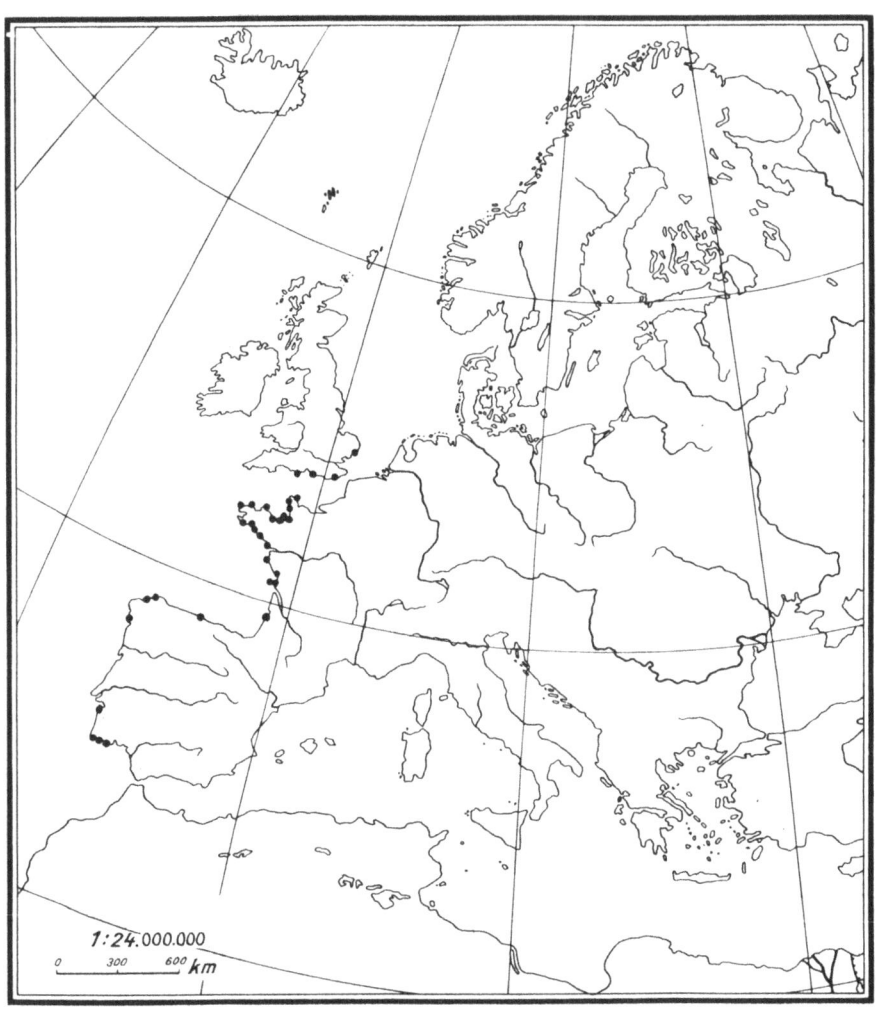

Carte n°1 : *Puccinellio maritimae-Arthrocnemetum perennis typicum*

Tableau n° 2: *Puccinellio maritimae – Arthrocnemetum perennis typicum*

	1	2	3	4	5	6	7	8	9	10	11	12	13	14	15	16	17	18	19	20	21	22	23	24	25	26
Numéros des relevés:	1	2	3	4	5	6	7	8	9	10	11	12	13	14	15	16	17	18	19	20	21	22	23	24	25	26
Recouvrement (en %):	90	70	70	80	80	80	60	70	70	70	90	50	40	25	25	50	70	70	80	90	40	80	50	80		60
Surface (en m²):	1	0,5	2	2	2	10	0,25	1	1	1	50	2	2	10	10	5	1	5	1	10	2	1	20	5	6	2
Nombre de relevés:																										
Nombre d'espèces:	2	2	2	2	2	3	3	3	3	3	3	3	3	3	3	4	4	4	4	5	5	5	5	5	6	6
Caractéristiques d'association:																										
Arthrocnemum perenne	54	32	44	43	34	44	44	43	54	55	43	43	33	32	32	33	44	33	33	43	32	54	43	44	34	32
Puccinellia maritima															+2	22				23			+(+2)	+(+2)	23	+
Bostrychia scorpioides						23													12	24		12				
Différentielles de variantes:																										
Salicornia stricta et cf.	11	22	i	21	23	12	22	21	11	+	23	11	+	11	11	22	+	21								
Halimione portulacoides																			23	+2	11	+	11	21	+	21
Compagnes:																										
Suaeda maritima							+				12					+2	22	+		+2	21	+		22		21
Aster tripolium								+2	+2	+				+				12			21		+2	34		
Spergularia media																		+	23		+2	+2				
Limonium vulgare																	12				12°	+2	+2 (i)		12	
Triglochin maritimum																									12	
Accidentelles:												1	1								1				2	

38

Tableau n° 2 (suite)

	27	28	29	30	31	32	33	34	35	36	37	38	39	40	41	42	43	44	45	46	47	48	49	50	51	52
Numéros des relevés:	27	28	29	30	31	32	33	34	35	36	37	38	39	40	41	42	43	44	45	46	47	48	49	50	51	52
Recouvrement (en %):	100	80	60	90	90	80	75	70	80	90	80	80	80	100	90	90	90	85	70	90		80	80	80	90	
Surface (en m²):	2	4	2	10	10	5	10	10	5	5	2	2	1	2	5	2	1	3	10	10	2	10	10	10	2	3
Nombre de relevés:	6	5	5	7																						
Nombre d'espèces:	6	5	5	7	1	1	1	1	1	1	1	1	1	1	1	1	1	2	2	2	2	2	2	2	2	3
Caractéristiques d'association:																										
Arthrocnemum perenne	54	43	44	43	55	55	55	32	33	54	45	45	55	53	55	55	54	54	44	54	54	54	44	54	55	44
Puccinellia maritima	23	+	+2	22														12		12						22
Bostrychia scorpioides			23	23																	+					
Différentielles de variantes:																										
Salicornia stricta et cf.	+	33	33	22																						
Halimione portulacoides		+°	+2	+																						
Compagnes:																										
Suaeda maritima	+			+															12							
Aster tripolium	+2																					i	+	23	11	
Spergularia media		11	11	11																						
Limonium vulgare	+																									
Triglochin maritimum																										12
Accidentelles:																										

39

Tableau n° 2 (suite)

	53	54	55	56	57	58	59	60	61	62	63	64	65	66	67	68	69	70	71	72	73	74	75	76	77	78
Numéros des relevés:	53	54	55	56	57	58	59	60	61	62	63	64	65	66	67	68	69	70	71	72	73	74	75	76	77	78
Recouvrement (en %):	70	95	60	75	75	50	100	80		70	50	90	75	80	40	90	90	100		90	100	100	90	90	60	50
Surface (en m²):	10	10	1	1	1	1	5	10	4	1	4	5	1	10	5	5	2	2	3	10	4	5	5	1	2	1
Nombre de relevés:	3	3	3	3	3	3	3	3	4	4	4	4	4	2	2	2	2	3	3	3	3	3	3	3	3	3
Nombre d'espèces:	3	3	3	3	3	3	3	3	4	4	4	4	4	6	2	2	2	2	3	3	3	3	3	3	3	3
Caractéristiques d'association:																										
Arthrocnemum perenne	43	44	43	44	33	33	43	43	33	44	43	43	34	44	34	52	55	55	44	44	54	44	54	44	43	44
Puccinellia maritima		33		22				23	+2	12	+	22	24	+2												
Bostrychia scorpioides	33					+													+	12			12	12	+	33
Différentielles de variantes:																										
Salicornia stricta et cf.														+												
Halimione portulacoides														+2	12	22	+2	+	23	23	24	23	22	32	+	12
Compagnes:																										
Suaeda maritima	11													+												
Aster tripolium	12		+2	+2			+	32	+	+	+		11				+									
Spergularia media					22	12			+2	+	+	33	+2	22												
Limonium vulgare		+			33																					
Triglochin maritimum												21														
Accidentelles:		1																			1	1				

Tableau n° 2 (suite)

Numéros des relevés:	79	80	81	82	83	84	85	86	87	88	89	90	91	92	93	94	95	96	97	98	99	100	101	102	103	104	105	Synth.
Recouvrement (en %):	50	75	100	85	80	80	90	75	90	90	90	60	1	100	100	50	90	90	80	70	100	100	100	90	90	90	60	
Surface (en m²):	1	1	6	2	5	5	5	5	10	1	5	1	0,5	4	4	20	1	1	10	10	15	5	5	1	1	3	10	3,3
Nombre de relevés:																												105
Nombre d'espèces:	3	3	3	3	3	3	3	4	4	4	4	4	4	4	4	4	4	5	5	5	5	5	6	6	6	6	6	
Caractéristiques d'association:																												
Arthrocnemum perenne	32	43	44	44	44	43	43	54	44	45	54	43	54	54	44	22	43	44	54	43	55	43	34	44	54	54	44	V
Puccinellia maritima			44	44					21	11	+2		+	+2	+	+	22			+2	22	+	22	+	+	12	12	II
Bostrychia scorpioides	23	44	44	44	+	23	54	23	23		23	23		33	33		32	33	12	23	23	23	12	+2	+	11	23	II
Différentielles de variantes:																												
Salicornia stricta et cf.										12	21	11				22							23	33				II
Halimione portulacoides	+	22	22	23	+2	+	21	22	22				+2	+	+		32	32	+2	22	+2	+2			12	+2	+2	III
Compagnes:																												
Suaeda maritim		+											+°					22	+		22		12			12		II
Aster tripolium																			+		+2	23		+	11			II
Spergularia media																			23				+2		+	+		I
Limonium vulgare								12		+2		+								23			12	+			12	I
Triglochin maritima																43												+
Accidentelles:																								1				

Légende du tableau n° 1 et de la carte n° 1

PORTUGAL:
rel. n° 1, 35 à 38, 55 (*Scirpus maritimus* +2), 67 à 69, 77 à 79, 90: Faro – Portimao – Lagos – Rio Sado.

ESPAGNE:
rel. n° 25 (*Spergularia marina* 11, *Salicornia ramosissima* 1,2), 31, 32, 52: Plaja de Lanjuda. – n° 33, 86: La Coruña. – n° 34, 44: Cedeira. – n° 45, 53: St-Vincente de la Barquera.

FRANCE:
rel. n° 3, 17, 56, 57, 61, 63, 70, 71, 96: Arcachon (33). – n° 8, 9, 10: Ile d'Oléron (17). – n° 94: Ile Madame (17). – n° 11, 99: La Faute/mer (85). – n° 40: Ile de Noirmoutier (85). – n° 4, 12 (*Puccinellia distans* +), 13 (*Puccinellia distans* 2,1), 18, 22, 47, 64, 100: Pembron (44). – n° 58: Locmariaquer (56). – n° 59: Plouharnel (*Scirpus pungens* 2.1) (56). – n° 62: Lanester (56). – n° 75, 89, 98: Fouesnant (29). – n° 6, 16, 20, 46, 54, 72, 73 (*Juncus maritimus* 11), 74 (*Juncus maritimus* +), 88, 101, 102: Ile Chevalier (29). – n° 28 à 30, 87: Le Conquet (29). – n° 66, 97: Locquénolé (29). – n° 76, 103: Tréguier (22). – n° 85: Talbert (22). – n° 84: Yffiniac (22). – n° 23: Hillion (22). – n° 5: Le Frémur (22). – n° 42, 51, 60, 104 (*Salicornia ramosissima*): La Fresnaye (22). – n° 48 à 50: Le Guildo (22). – n° 24: St-Jouan (35). – n° 19: La Ville Ger (35). – n° 41: St-Benoit des Ondes (35). – n° 65: Cherrueix (35). – n° 14, 15, 27: Lessay (50). – n° 105: Barneville (50). – n° 39, 43, 81, 82, 92, 93, 95: St-Vast-La-Hougue (50).

ANGLETERRE:
rel. n° 2, 80: Wittering. – n° 7: Pagham Harbour. – n° 91: Rye. – n° 21 (*Limonium humile* +): Blackeney Point.

Bostrychia scorpioides, malgré sa présence variable possède aussi une réelle valeur différentielle pour l'association atlantique. Ajoutons que son individualisation bénéficie aussi de l'originalité des contacts.

Variation de l'association: Sur la base de quelques 181 relevés répartis en 4 tableaux, il nous paraît possible de distinguer actuellement quatre sous-associations:

— Sous-association typique

Puccinellio maritimae-Arthrocnemetum perennis typicum Géhu 1975 em. Tableau n° 2: (105 relevés)

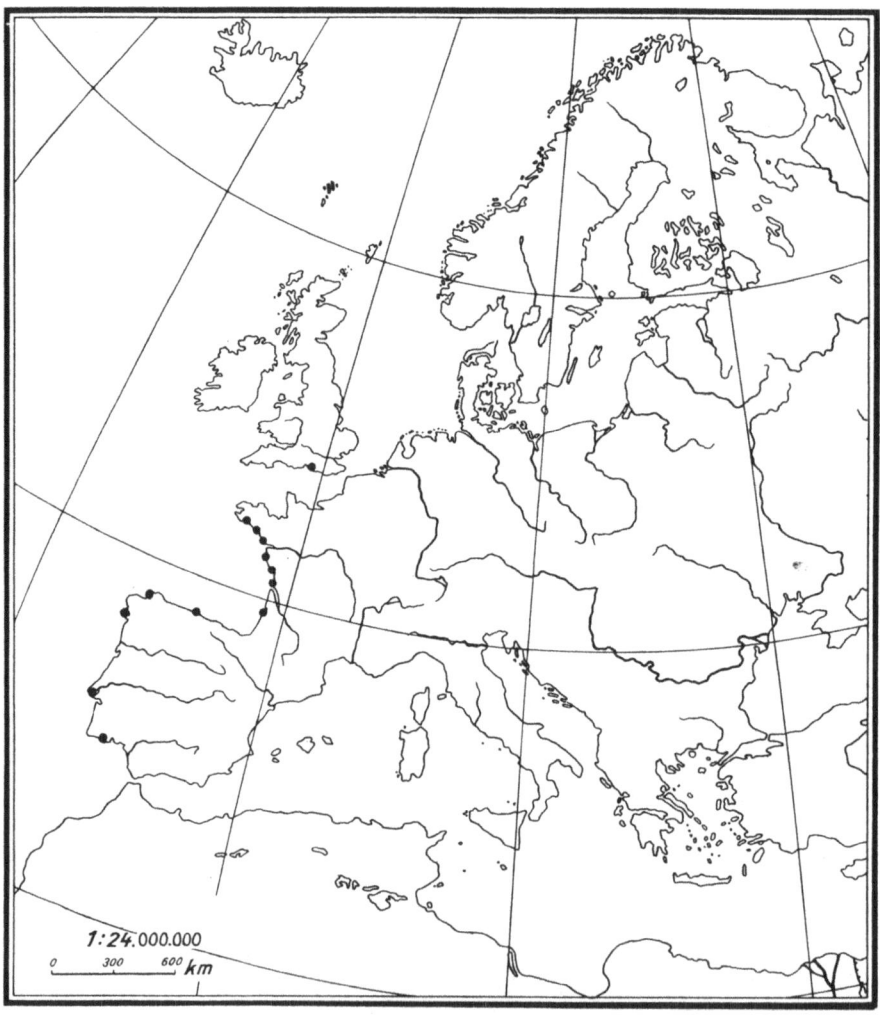

Carte n°2 :*Puccinellio maritimae-Arthrocnemetum perennis spartinetosum maritimae*

Typus nominis de l'association et de la sous-association: rel. n° 62 du tableau n° 2. Carte de répartition n° 1.

La sous-association est présenté dans toute l'aire du groupement. Elle possède trois variantes, l'une inférieure à Salicornes annuelles (type *stricta* coll.), l'une moyenne, et la 3ème supérieure à *Halimione portulacoides*. Sur la base du matériel, actuellement disponible, il nous paraît préférable en effet de réduire les sous-associations *salicornietosum* et *halimionetosum* de 1975 à de simples variantes de contact inférieur ou supérieur.

— Sous-association *spartinetosum maritimae*

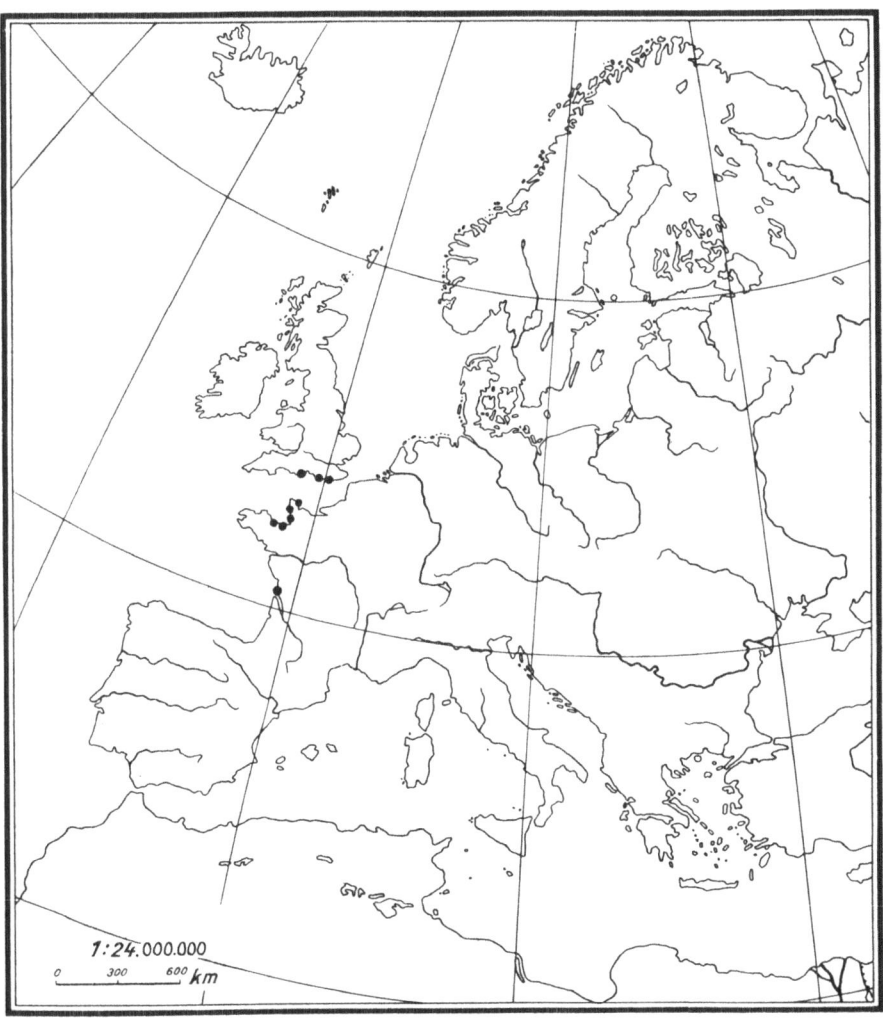

Carte n°3 : *Puccinellio maritimae-Arthrocnemetum perennis spartinetosum townsendii*

43

Puccinellio maritimae-Arthrocnemetum perennis spartinetosum maritimae Fontes 1945 em. J.M. et J. Géhu 1977. Tableau n° 3: (29 relevés)

Typus nominis de la sous-association: in J.M. et J. Géhu 1977. Carte de répartition n° 2.

La sous-association bien différentiée par *Spartina maritima* correspond aux niveaux inférieurs du groupement dans la partie occidentale de son aire où subsiste *Spartina maritima.*

— Sous-association *spartinetosum townsendii*

Puccinellio maritimae-Arthrocnemetum perennis spartinetosum townsendii ss. ass. nov. Tableau n° 4: (35 relevés)

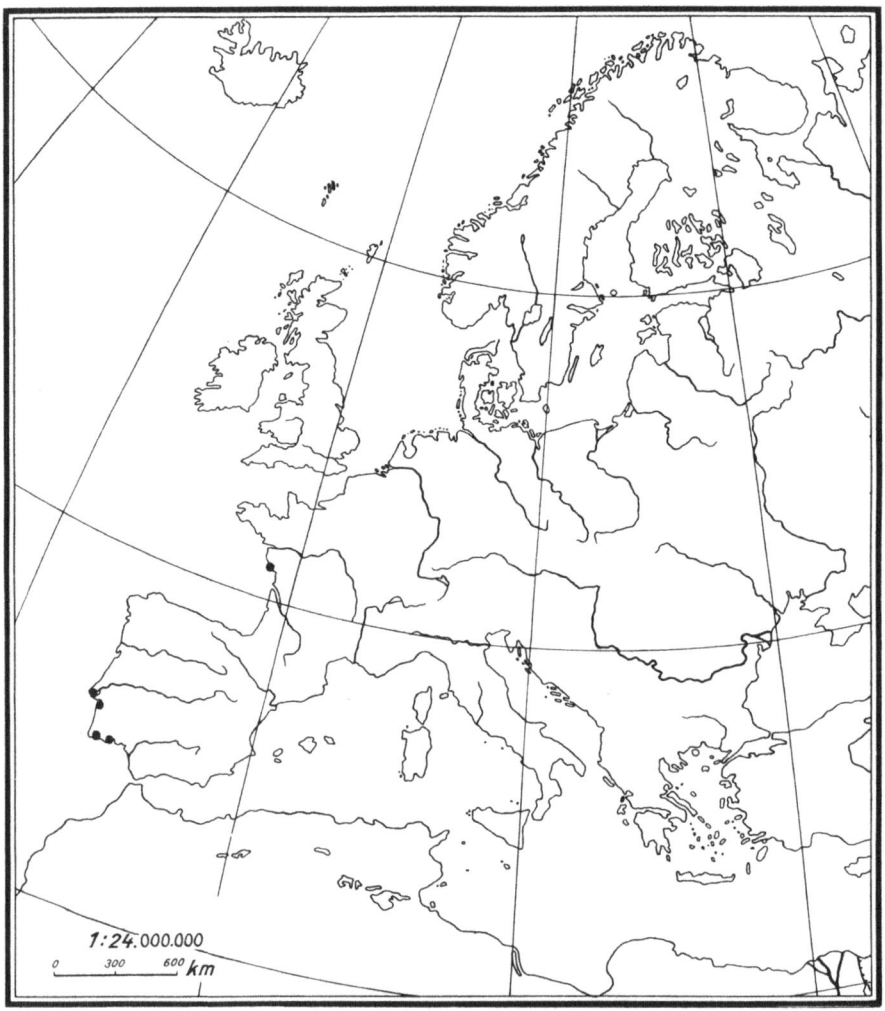

Carte n°4 : *Puccinellio maritimae-Arthrocnemetum perennis arthrocnemetosum fruticosi*

Tableau n°: 3   *Puccinellio maritimae-Arthrocnemetum perennis spartinetosum maritimae*

	1	2	3	4	5	6	7	8	9	10	11	12	13	14	15	16	17	18	19	20	21	22	23	24	25	26	27	28	29	
Numéros des relevés:	1	2	3	4	5	6	7	8	9	10	11	12	13	14	15	16	17	18	19	20	21	22	23	24	25	26	27	28	29	
Recouvrement (en %):	80	75	70	80	80	50	90	95	80	85	100	70	100	80	75	85	90	100	80	90	90	95	80	80	100	90	60	80	80	
Surface (en m²):	1	1	10	1	2	5	5	1	5	5	5	10	10	4	6	1	2	5	1	10	5	5	5	5	1	2	5	5	7	
Nombre de relevés:																														29
Nombre d'espèces:	2	3	3	3	3	3	3	3	3	3	4	4	4	4	5	5	5	6	6	4	4	5	5	5	6	6	6	6	7	4,3
Caract. et Diff. d'association:																														
Arthrocnemum perenne	55	32	43	34	44	34	54	34	43	44	44	54	43	32	54	44	44	44	43	44	44	45	23	43	33	55	54	54	23	V
Puccinellia maritima	+	+	21			12		12					32	21	11	+	32	23	11			24	22			+	12	23	21	III
Bostrychia scorpioides				23	+				11	+	33		23				12	43	11	33				12	+2		23	+		III
Diff. de ss.-assoc.:																														
Spartina maritima	+2	22	21	+2	23	21	22	22	32	21	22	11	11	21	+	11	11	12	11	21	21	+	32	22	+	+	21	21	32	V
Diff. de variante:																														
Halimione portulacoides																				+	11	12	12	32	34	12	11	+	12	II
Compagnes:																														
Aster tripolium			+				+									+									+	+2	+			II
Suaeda maritima														+		+	22	+	+		21	11	+							II
Limonium vulgare												i												22		12		+		I
Spergularia media											+2	21			+			23												I
Accidentelles:																			1											

Légende du tableau n° 3 et de la carte n° 2

PORTUGAL:
rel. n° 22, 23, 29 (*Salicornia stricta* +): Sacavem.

ESPAGNE:
rel. n° 11, 18: Plaja de Lanjuda. – n° 2, 15 (*salicornia stricta* +): La Coruña. – n° 3, 12: St-Vincente de la Barquera.

FRANCE:
rel. n° 4, 25, 26: Arcachon (33). – n° 7, 20, 27: Marennes (17). – n° 1: La Faute/Mer (85). – n° 8, 16 (*Salicornia stricta* +): Baie de Bourgneuf (85). – n° 19 (*Salicornia stricta* +): Baie de Bourgneuf (85). – n° 24: Pembron (44). – n° 17: Noyalo (56). – n° 9, 28: Benance (56). – n° 10: Plouharnel (56). – n° 5, 6: Fouesnant (29). – n° 13: Ile Chevalier (29).

ANGLETERRE:
rel. n° 14, 21: Wittering.

Tableau n° 4 : *Puccinellio maritimae-Arthrocnemetum perennis spartinetosum townsendii*

	1	2	3	4	5	6	7	8	9	10	11	12	13	14	15	16	17	18	19	20	21	22	23	24	25	26	27	28	29	30	31	32	33	34	35	
Numéros des relevés:	1	2	3	4	5	6	7	8	9	10	11	12	13	14	15	16	17	18	19	20	21	22	23	24	25	26	27	28	29	30	31	32	33	34	35	35
Recouvrement (en %):	80	80	60	60	90	80	50	60	70	80	80		80	95		65	75	50	85	85	80	90	10	50	100	75	50	50	70	70		80	80	100	60	
Surface (en m²):	1	2	1	1	2	2	2	2	3	3	10		5	1		2	1	5	2	2	5	5	2	5	2	1	10	2	2	30		1	5	2	2	
Nombre de relevés:																																				35
Nombre d'espèces:	3	3	4	4	4	5	7	2	2	2	2	2	2	2	3	3	3	3	3	3	4	4	4	4	4	4	5	3	3	4	4	5	5	6	6	3,9
Caract. et diff. d'association:																																				
Arthrocnemum perenne	32	44	33	44	55	54	44	44	44	54	54	44	44	54	55	45	45	43	54	54	43	43	44	43	55	43	34	43	44	34	44	54	54	44	43	V
Puccinellia maritima	44		22	12		+2	+	44	44	+	+	+2						+		+	13	22		+2	+	12	+2				22	+2	+2	21	+2	III
Bostrychia scorpioides																12																		12		+
Esp. diff. de ss-assoc.:																																				
Spartina townsendii	22	+	+	12	+	+	21	21	+2	+	+2	22	22	+	+	+	+	+	+	+2	32	+	+2	+	+	21	+2	12°	12	11	21	12	12	+2	12	V
Esp. diff. de variantes:																																				
Salicornia stricta	11	23	11	11			21																													I
Halimione portulacoides																											+2	+2	21	12	13	11	+2	+2		II
Compagnes:																																				
Aster tripolium																	23		+	+	+	+	12	+	+	22	32				i	i				II
Sueda maritima					+2	+																														I
Limonium vulgare							+2																				+									+
Accidentelles:							2																													

Légende du tableau n° 4 et de la carte n° 3

FRANCE:
Rel. n° 29: Ile d'Oléron (17). – n° 22: Yffiniac (22). – n° 34: Hillion (22). – n° 18, 24: Sables d'Or les Pins (22). – n° 2, 12, 15, 16, 17: Le Frémur (22). – n° 10, 11: Le Guildo (22). – n° 13, 33: St-Benoît des Ondes (35). n° 19, 20, 21: Le Vivier (35). – n° 28: Lingreville (50). – n° 5, 25: Lessay (50). – n° 14, 23, 26, 27, 35: Barneville (50). – n° 8, 9: St-Vast-La-Hougue (50).

ANGLETERRE:
Rel. n° 7 (*Spergularia media* +2, *Triglochin maritimus* +): Swanage. – n° 1, 3, 4, 6, 31: Wittering. – n° 30, 32: Rye.

Tableau n° 5: *Puccinellio maritimae-Arthrocnemetum perennis arthrocnemetosum fruticosi*

Numéros des relevés:	1	2	3	4	5	6	7	8	9	10	11	12	
Recouvrement (en %):	50	9	80	80	40	100	80	100	100	80	80	80	
Surface (en m²):	2	5			5		2			15			
Nombre de relevés:													12
Nombre d'espèces:	2	2	3	3	3	4	4	5	5	5	6	6	4
Carac. et diff. d'association:													
Arthrocnemum perenne	44	55	45	34	34	34	44	34	33	45	44	45	V
Puccinellia maritima						34		34	34	12	13	12	III
Bostrychia scorpioides							+						+
Esp. diff. de ss.-assoc.:													
Arthrocnemum fruticosum	12	12	+	11	12	12	+2	12	11	13	+	12	V
Esp. unités supérieures:													
Halimione portulacoides			+2	22	12		+2	12	12	+2	+2	+	IV
Compagnes:													
Aster tripolium						+		+	+				II
Suaeda maritima										+	+		I
Salicornia stricta											+	12	I
Arthrocnemum glaucum												11	+

Légende du tableau n° 5 et de la carte n° 4

PORTUGAL:
Rel. n° 1, 2, 5, 7: Sud Portugal
n° 3, 4, 6, 8, 9, 11, 12: Sacavem

FRANCE:
Rel. n° 10: La Faute/Mer (85).

Typus nominis de la sous-association: rel. n° 22 du tableau n° 4. Carte de répartition n° 3.

C'est une sous-association inférieure 'néophyte' liée au développement du *Spartinetum townsendii* dans la partie nord occidentale de l'aire du *Puccinellio-Arthrocnemetum perennis.*

— Sous-association *arthrocnemetosum fruticosi*

Puccinellio maritimae – *Arthrocnemetum perennis arthrocnemetosum fruticosi* J.M. et J. Géhu 1977. Tableau n° 5: (12 relevés)

Typus nominis de la sous-association: in J.M. et J. Géhu 1977. Carte de répartition n° 4.

C'est la sous-association des niveaux les plus élevés dans la partie la plus thermophile de l'aire (Sud-Ouest ibérique et Centre Ouest français).

Bibliographie

Braun-Blanquet, J., 1933. *Ammophiletalia* et *Salicornietalia* méditerranéens. *Prodrome des groupements végétaux*. Fasc. 1, 23 p. Montpellier.

Corillion, R., 1953. Les halipèdes du Nord de la Bretagne, études phytosociologique et phytogéographique. *Rev. Gen. Bot.*, 60, 124 p. Paris.

Fontes, F.C., 1945. Algunas características fitosociológicas dos 'Salgados' de Sácavem. *Bol. da Sociedade Broteriana* 2eS. 19 (2), 789-813. Coimbra.

Géhu, J.M., 1973. Sur la signification écologique et dynamique et la vicarance géographique des groupements à *Halimione portulacoides* des côtes atlantiques européennes. Ber. d. Intern. Symp. d. Intern. Verein. f. Vegetations- u. Sukzessionsforschung 53-70. Rinteln (paru 1975).

Géhu, J.M., 1975. Approche phytosociologique synthétique de la végétation des vases salées du littoral atlantique français (synsystématique et synchorologie). Colloques phytosociologiques IV, 395-462. Lille (paru en 1976).

Géhu, J.M., 1975. Essai systématique et chorologique sur les principales associations végétales du littoral atlantique français. *Ann. Real Acad. Farmacia*, 41 (2): 207-227. Madrid.

Géhu, J.M. et Delzenne, CH, 1975. Apport à la connaissance phytosociologique des prairies salées de l'Angleterre. *Colloques phytosociologiques* IV, 227-248. Lille (paru 1976).

Géhu, J.M., Foucault, B. de & J. Géhu-Franck, 1978. Les végétations à *Arthrocnemum fruticosum* du littoral atlantique français. *Bull. Soc. Bot. Nord France*, 31, (sous presse), Lille.

Géhu, J.M. et J. Géhu-Franck, 1977. Quelques données sur les *Arthrocnemetea fruticosi* ibériques sud-occidentaux. *Acta Botanica Malacitana*, 3, 145-157. Malaga.

Address of the Authors:
Prof. Dr. J.-M. Géhu,
Station de Phytosociologie, 59270 Boilleul
et Faculté de Pharmacie, Laboratoire de Botanique, 59045 Lille, Cédex,
France

L'ETAGEMENT ALTITUDINAL DE LA VEGETATION AU CHILI CENTRAL: LES PROFILS PHYTOGEOGRAPHIQUES

Victor G. Quintanilla

Abstract

Central Chile extends under a mediterranean type of climatic influence between the 30th and 38th grades of southern latitude. Due to a great concentration of population and the agricultural activity this part of the national territory practically is lacking in natural vegetation in the interiors valleys and coastal sectors of the precordillera.

Therefore, a descriptive analisis of the vegetation is more realistic if we do it in the Andes and in the coast ranges, following the criteria of altitudinal zonalization.

The northern sector of the region shows in its vegetation levels a certain similarity in its composition and physionomy with the natural formations of lowlands and plain regions which belong basically to latitudinal grouping typical of the region.

However, south of the 33 latitude, this similarity begins to disappear mainly due to high humidity registered in the mountains. The sclerophytal woods are replaced by mixed formations where the predominant species are the *Fagaceae*, *Lauraceae* and conifer.

Between the 37 and 38 grades of southern latitude, in both mountains ranges an almost pure formation of *Araucaria araucana* woods, an endemic conifer of South America is developed. In the coastal range, this formation reachs its optimun ecology between 800 and 1300 meters aproximately while in the Andes this one is achieved only at 1.600 meters of altitude.

The altitudinal descent of both ranges is almost parallely with the latitudinal development to the south and this is one of the reasons why some trees have a wide distribution in some altitudinal levels. (*Nothofagus obliqua*, *Nothofagus alpina*, *Nothofagus dombeyi*).

It is worth noting that at present, the anthropologycal action and the reforestation with foreign species are seriously affecting the preservation of the natural woods on this mountain ranges.

La région considérée pour nous à l'intérieur du Chili Central, s'étend d'une manière approximative entre les 30 et 38 degrés de latitude sud. Au point de vue climatique cette région participe en général des caractères d'un climat avec tendance méditerranéenne qui présente une longue période de sécheresse. Cette sécheresse diminue à mesure qu'on avance en latitude vers le sud, comme conséquence de l'augmentation de l'humidité et de la modération des températures qu'on observe particulièrement sur les versants des cordillères (Fig. 1).

Au point de vue physiographiques, la partie septentrionale connue régionalement comme le 'Norte Chico' (Petit Nord), a une orographie accidentée avec de chaînes de montagnes intérieures qui se distribuent transversalement depuis la Cordillère des Andes pour l'Est (hauteur moyenne du massif andin proche au 5.000 mètres, vers la Cordillère de la Côte pour l'Ouest, laquelle a des hauteurs maximums de l'ordre des 1.500 mètres. Ces chaînes transversales permettent l'existence d'un réseau embrouillé de bassins hydrographiques dans lequel se dé-

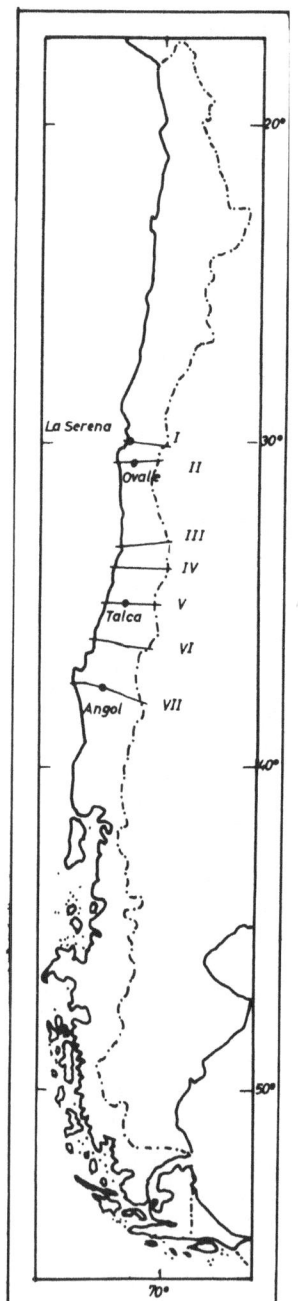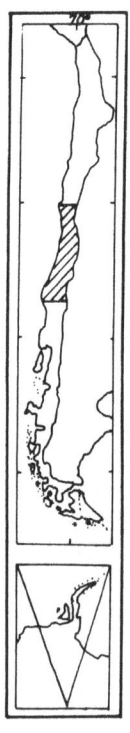

FIG. 1 - SITUATION DE LA ZONE D'ETUDE ET DES PROFILS PHYTOGEOGRAPHIQUES
AU CHILI CENTRAL.

50

veloppe particulièrement l'habitat urbaine et rural. Le Norte Chico a un climat méditerranéen accentué avec des précipitations qui ne dépassent les 200 mm et avec une longue période de sécheresse annuelle qu'oscille entre 8 et 10 mois. Telles caractéristiques contribuent à donner à la végétation un aspect éminemment sémiaride.

Au sud des 33° depuis Quintero et Los Andes, la physiographie et le climat du pays ont quelques changements importants. Les vallées et chaînes transversales qui caractérisaient le Norte Chico finissent dans la fosse du fleuve Aconcagua, et à partir d'ici jusqu'au sud de notre zone d'étude le relief prend une physionomie longitudinale tripartite. Entre la Cordillère des Andes et la Cordillère de la Côte, se développe une grande vallée ou déprétion centrale d'une largeur moyenne de 80 ou 90 kms. approsimativement. Dans la latitude 33° Sud, les sommets supérieur des Andes arrivent aux 6.000 mts. pour descendre ensuite peu à peu vers le sud et avoir autour de 3.000 mts. vers les 38° S. On a aussi un phénomène similaire, même avec des hauteurs inférieurs aux 2.500 mts. dans la Cordillère de la Côte, où la chaîne se coupe dans quelques secteurs mettant en contact les plaines littorales avec la Vallée Longitudinale.

Dans cette parti du Chili le climat méditerranéen est moins accentué, avec un accroissement important des précipitations depuis Santiago vers le sud et on observe aussi une diminution modérée de la période de sécheresse (Quintanilla 1974). Nous prions se rapporter aux diagrammes ombrothermiques correspondants, suivant Gaussen-Walther, qui accompagnent aux profils phytogéographiques.

D'autre part, la période pluvieuse et le total des précipitations sont plus élevés dans la versant pacifique de la cordillère côtière et aussi dans la précordillère andine. Toutes ces conditions ont une influence sur le développement de la végétation, laquelle au Sud du 33° peut présenter des formations arborescentes mésophytes et encore de caractère higrophyte dans les secteurs les plus humides des cordillères.

1. Caractéristiques géobotaniques du Chili Central

La plupart des études biogéographiques sur le Chili, coincident à distinguer un ensemble de formations sémiarides évidencié par une dominance du buisson ou du 'matorral' xérophyte pour ces aires du territoire situées entre les 30 et 33° latitude sud. Après, à peu près jusqu'aux 38° S., se développe la formation de la forêt sclérophylle laquelle est remplacée à cause de l'avancement latitudinale par des composants de la forêt caduque et humide, typique du sud du pays, sur les versants les plus hauts et les plus humides des deux cordillères. Il faut remarquer, que des conditions écologiques locales de plus grande prédominance d'humidité ou de sécheresse; permettent l'existence des communautés végétales un peu azonaux à l'intérieur de ces trois régions phytogéographiques. De cette manière, vers le littoral de la région sémiaride (entre La Serena et Quintero) se développent d'intéressants groupements végétaux de type hygrophile dont on doit leur durée principalement à la présence pendant toute l'année, de brouillards côtiers. Exemple typique, c'est la forêt des brumes du Fray Jorge vers les 30° 30' lat. Sud.

Dans la Vallée Longitudinale, la forêt sclérophylle a disparu presque toute à cause de l'intervention humaine, en donnant lieu à une formation steppique avec un arbuste sémixérophyte dominant qui colonise aujourd'hui une grande partie d'endroits du Chili Central avec des conditions écologiques médiocres. Il s'agit de l'*Acacia caven*, dont ses groupements constituent la formation de l' 'Espinal' laquelle vers le sud du 36° prend la physionomie d'une seudosavane, trouvant actuellement sa limite méridionale d'expansion autour des 39° Sud.

Pourtant, notre zone d'étude est fondamentalement insérée dans la formation du forêt sclérophylle – Hartlaubgebiet d'après la classification régionale de la végétation chilienne proposée par Schmithüsen (1956) – et limitée vers le nord et le sud par deux régions végétationales différentes comme conséquence aussi des conditions climatiques très diverses (Fig. 2).

Au nord du 33° le régime médiocre des précipitations et la forte radiation solaire, déterminent fondamentalement l'existence d'une formation sémiaride d'arbustes épineux et succulents (Zwergstrauchgebietes des Kleinen Nordens, de Schmithüsen). Mais dans le sud du pays, plus au moins à partir des 37° S., les conditions bioclimatiques sont plus favorables pour les plantes, grâce aux lesquelles des formations tempérés d'arbres caduques se développent – avec dominance de *Nothofagus* – et nommée Gebiet des temperierten Sommerwaldes der gemässigten zone pour Schmithüsen. Encore, dans les secteurs où le régime de précipitation et d'humidité est presque constant la 'pluviselva valdiviana' se développe (le Gebiet der immer grünen Regenwälder).

2. Sur la convenance de l'utilisation des profils dans l'analyse descriptif de la végétation chilienne

Jusqu'aujourd'hui, les études et travaux biogéographiques du Chili ont été orientées fondamentalement à donner une vision latitudinale et régionale, en grande partie déterminée par l'orographie et la structure géographique du territoire. De cette façon on fait des études qui se superposent des bandes verticales latitudinales du pays en mettant en relation entre eux les types de climats, de sols et de végétation. Par contre on a effectué très peu d'études horizontaux qui travaillent avec des profils ou transects phytogéographiques depuis le littoral jusqu'à la cordillère des Andes et dans lesquelles on peut observer un numéro plus important de zones de vie spécifiques tandis; qu'il est beaucoup plus difficile d'obtenir des résultats valables sur d'autres parties du territoire, à travers d'une conception latitudinale régionale.

Nous pensons que le deux méthodes sont importantes et nécessaires pour la description spatiale de la végétation dans un pays assez étendue comme le Chili, dans lequel ses écosysthèmes sont aussi par l'orographie très influencés. Aujourd'-

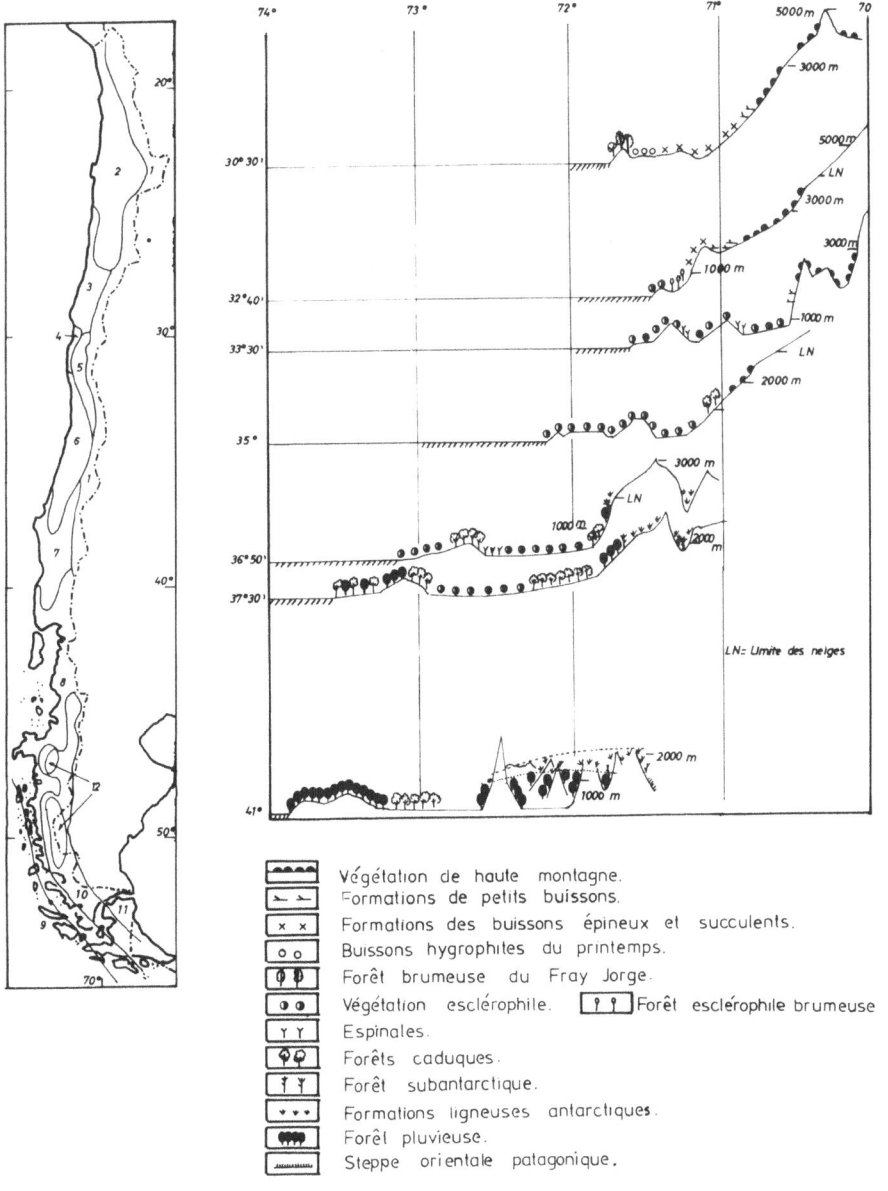

Végétation de haute montagne.	
Formations de petits buissons.	
Formations des buissons épineux et succulents.	
Buissons hygrophites du printemps.	
Forêt brumeuse du Fray Jorge.	
Végétation esclérophile. Forêt esclérophile brumeuse	
Espinales.	
Forêts caduques.	
Forêt subantarctique.	
Formations ligneuses antarctiques.	
Forêt pluvieuse.	
Steppe orientale patagonique.	

LN= Limite des neiges

1.- Formations andines 2.- Désert 3.- Formations subtropicaux des buissons rampant du Norte Chico
4.- District de La Serena 5.- Formations subtropicaux des succulents et buissons épineux du Norte Chico
6.- Formations esclérophiles 7.- Forêts caduques tempérés 8.- Formations des forêts sempervirents pluvieuses 9.- Toundras subantarctiques 10.- Forêt caduque subantarctique
11.- Steppes patagoniques 12.- Andes Méridionaux.

Fig 2 - Profils altitudinales de la végétation du Chili Central.

Distribution régional de la végétation chilienne.

[d'aprés J. Schmithüsen, 1956]

53

hui, pour le Chili Central, l'étude de la distribution altitudinale des communautés végétales est beaucoup plus réaliste et objectif, que celle de type régional. Nous avons deux raisons fondamentales, pour soutenir cette thèse.

Comme dans toutes les zones méditerranéennes du globe, la région chilienne pertinente a expérimenté, l'impact humain depuis longtemps. Donc l'altération des paysages naturelles de la Vallée Longitudinale et des versants précordillerains, particulièrement celles de la chaîne du littoral, a été très intensive. Dans quelques secteurs ce phénomène est irréversible empêchant la régénération de la végétation naturelle, en favorisant de cette manière la colonisation des plantes étrangères avec une plus grande tolérance écologique, ou en donnant origine aux procès érosifs avancés, qui ont obligé à reboiser avec des espèces exotiques. De cette façon on explique en partie aussi, la grande expantion sur la Vallée Longitudinale de l' 'espino', *Acacia caven*, en dépit des aires de distribution des forêts sclérophylles et lauripholiés. C'est a dire aujourd'hui, à exception de petits secteurs, il ne reste plus de végétation naturelle primitive dans cette zone physiographique du pays.

Une autre raison est de caractère climatique. La 'désert avance' est une asévération que depuis beaucoup d'années on a entendu au Chili. Les motifs qui l'expliquent sont de nature diverse. Quelquefois on a pensé les trouver dans l'action irrestricte de l'homme, qui à travers de ses multiples modes d'intervention a influencé sur les systèmes écologiques naturels; dans d'autres opportunités les raisons sont d'ordre géophysiques − surtout climatiques − argumentant le plupart des fois, qu'il s'agit de l'action en ensemble des deux éléments signalés.

En ce qui concerne la végétation naturelle, nous admettrons qu'il est extrêmement difficile de distinguer la cause des changements produits dans les communautés végétales. Quelque fois, ces altérations sont provoquées par l'action des facteurs anthropogéniques et biotiques, et finalement, nous trouvons avec celles qui ont l'origine dans les changements cumulatifs qui se produisent dans le mosaïque de microclimats, sur de grandes étendues de terrain, par une régression du tapis végétal à cause de l'action humaine vers un type de communautés plus aride (Hajek et autres, 1972).

Les altérations climatiques du Chili ont été de caractère cyclique comme il arrive dans d'autres parties du monde. Dans le dernier siècle elles ont eu une tendance à se présenter chaque 10 ou 40 ans. Cette sécheresse dans la zone de tendance méditerranéenne a eu et a des répercussions importantes. Particulièrement ici, il s'agit d'une zone qui présente nettement differenciés les périodes des pluies et l'absence de celles, là pour être la zone qui a la plus grande extension géographique et avec la plus grande partie de population du pays en incorporant aussi la plupart des aires d'utilisation agricole. D'autre part, l'existence d'un réseau approprié de postes météorologiques actuellement en service et avec une information de confiance, a permis d'effectuer des études de cette nature (di Castri-Hajek 1975).

Malheureusement au Chili nous n'avons pas encore une information assez complète sur les conditions de tolérance de beaucoup d'espèces naturelles ou introduites par l'homme bien que la flore et la faune chiliennes sont très labiles, ce qui permettra d'avancer avec une majeure précision dans un essai quelconque pour mesurer le degré de l'impact des changements climatiques sur celles-ci.

Quand on analyse pourtant, les variations climatiques qui ont foutté nos régions climatiques il est difficile d'établir si les facteurs qui ont eu une plus grande puissance dans la régression de la végétation chilienne, ont été l'action humaine ou le facteur climatique. Probablement on doit attribuer au climat la première importance dans ce phénomène.

Pourtant nous constatons que dans les dernières décades on a presque éliminé de la carte végétale du Chili quelques régions botaniques, desquelles on pense il serait très difficile de récuperer ou de reconstruire ses aires de dispertion géogra-

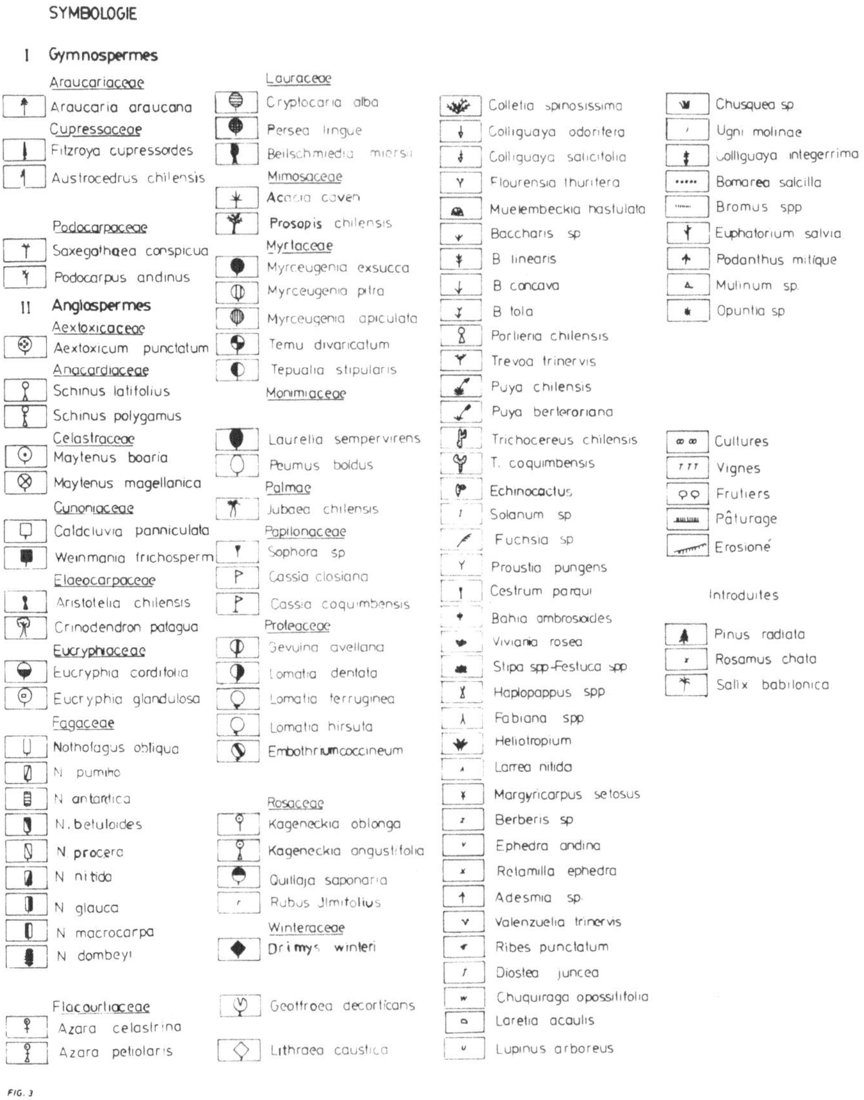

SYMBOLOGIE

I Gymnospermes

Araucariaceae
Araucaria araucana
Cupressaceae
Fitzroya cupressoides
Austrocedrus chilensis

Podocarpaceae
Saxegothaea conspicua
Podocarpus andinus

II Angiospermes
Aextoxicaceae
Aextoxicum punctatum
Anacardiaceae
Schinus latifolius
Schinus polygamus
Celastraceae
Maytenus boaria
Maytenus magellanica
Cunoniaceae
Caldcluvia panniculata
Weinmania trichosperma
Elaeocarpaceae
Aristotelia chilensis
Crinodendron patagua
Eucryphiaceae
Eucryphia cordifolia
Eucryphia glandulosa
Fagaceae
Nothofagus obliqua
N. pumilio
N. antartica
N. betuloides
N. procera
N. nitida
N. glauca
N. macrocarpa
N. dombeyi

Flacourtiaceae
Azara celastrina
Azara petiolaris

Lauraceae
Cryptocaria alba
Persea lingue
Beilschmiedia miersii
Mimosaceae
Acacia caven
Prosopis chilensis
Myrtaceae
Myrceugenia exsucca
Myrceugenia pitra
Myrceugenia apiculata
Temu divaricatum
Tepualia stipularis
Monimiaceae
Laurelia sempervirens
Peumus boldus
Palmae
Jubaea chilensis
Papilonaceae
Sophora sp
Cassia closiana
Cassia coquimbensis
Proteaceae
Gevuina avellana
Lomatia dentata
Lomatia terruginea
Lomatia hirsuta
Embothrium coccineum

Rosaceae
Kageneckia oblonga
Kageneckia angustifolia
Quillaja saponaria
Rubus Jimifolius
Winteraceae
Drimys winteri

Geoffroea decorticans
Lithraea caustica

Colletia spinosissima
Colliguaya odorifera
Colliguaya salicifolia
Flourensia thurifera
Muelembeckia hastulata
Baccharis sp
B. linearis
B. concava
B. tola
Porlieria chilensis
Trevoa trinervis
Puya chilensis
Puya berteroriana
Trichocereus chilensis
T. coquimbensis
Echinocactus
Solanum sp
Fuchsia sp
Proustia pungens
Cestrum parqui
Bahia ambrosioides
Viviania rosea
Stipa spp-Festuca spp
Haplopappus spp
Fabiana spp
Heliotropium
Larrea nitida
Margyricarpus setosus
Berberis sp
Ephedra andina
Retamilla ephedra
Adesmia sp
Valenzuelia trinervis
Ribes punctatum
Diostea juncea
Chuquiraga opossitifolia
Laretia acaulis
Lupinus arboreus

Chusquea sp
Ugni molinae
Colliguaya integerrima
Bomarea salcilla
Bromus spp
Euphatorium salvia
Podanthus mitique
Mulinum sp.
Opuntia sp

∞ ∞ Cultures
ттт Vignes
φ φ Fruitiers
Pâturage
Erosioné

Introduites
Pinus radiata
Rosamus chata
Salix babilonica

FIG. 3

55

phiques primitives. Probablement, nous serions alors en face d'un exemple typique de disclimax ou péniclimax.

Il y a une troisième raison qui appuie notre critérium. Il ne faut pas oublier que le Chili est un pays de pentes avec un territoire qui se déplace entre deux cordillères (en se souvient que les Andes sont unes des chaînes les plus hautes du globe), et quoiqu'elles possèdent une dépression intermédiaire étendue, la largeur moyenne ne dépasse pas les 100 kms., et jamais ne descend au niveau de la mer.

Pour ces raisons, nous pensons alors qu'une étude actuelle et réelle de la végétation dans la région méditerranéenne chilienne, est plus représentative quand on utilise un critérium descriptive altitudinale que quand on applique seulement un analyse régional. Actuellement, ce sont les versants des cordillères qui hébergent des groupements végétaux natifs représentatifs.

Pourtant, à partir de la classification régionale de la végétation établie par Schmithüsen, nous incorporons à son intérieur des profils altitudinaux de la végétation essayant à la fois une différentiation des étages végétaux. Quelques profils sont faits aux mêmes latitudes où le Professeur Schmithüsen a travaillé l'étagement de la végétation du Chili, et que nous avons agrandi postérieurement après les conclusions de nos parcours sur le terrain.

3. Les profils phytogéographiques du Chili Central

Dans d'autres opportunités nous avons expliqué le critère que nous avons appliqué pour différencier des étages végétaux au Chili Tempéré (Quintanilla 1974, 1977, 1978), sur la base de la composition physionomique des groupements représentés. Pourtant nous renvoyons au lecteur à ces travaux.

En relation avec les divisions géobotaniques régionales nous devons séparer nos profils de ceux correspondants à la zone sémiaride (30°-33° Sud), au Chili central méditerranéen typique (33°-36° Sud) et au Chili avec tendance méditerranéenne atenuée ou de type tempéré moderée (36°-38° Sud).

1. L'étagement au Norte Chico *(30° au 33° Sud)* Fig. 4 et 5

L'étage collinéen ou **basal,** a dans le Norte Chico un caractère typicament méditerranéen avec prédominance de xerophytes, mésophytes et arbres disperses thermophyles comme *Lithraea caustica* et *Schinus latifolius*. D'autre part, à cause de l'irrégulière orographie de la région, cet étage ne possède pas un paysage végétal homogéné. De même en raison des conditions bioclimatiques, les étages végétaux ont une composition floristique irrégulière en autre d'une distribution altitudinale très différenciée à cause de la diversité d'exposition des versants.

Evidemment a une prédominance des espèces adaptées à la sécheresse et à un déficit hydrique prolongé. (Nous savons que ici, la forêt higrophyle du Fray Jorge est une exception étant donné que fondamentalement elle s'agit d'un relicte climatique). Les Cactées sont remarquables (*Trichocereus coquimbensis* et *Echinocac-*

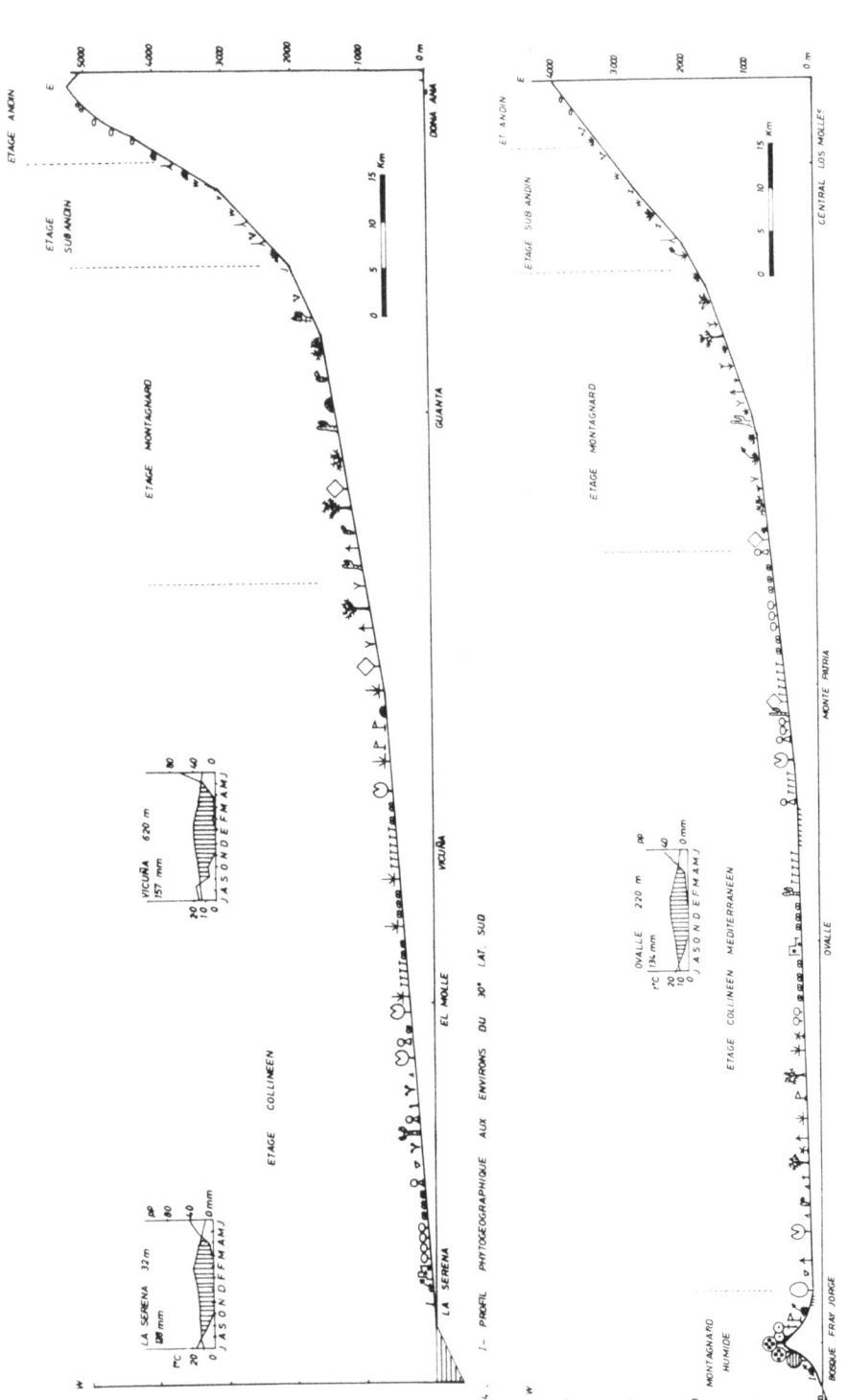

FIG. 4. 1.- PROFIL PHYTOGEOGRAPHIQUE AUX ENVIRONS DU 30° LAT. SUD

FIG. 5. 11 PROFIL PHYTOGEOGRAPHIQUE AUX ENVIRONS DU 30° 35' LAT. SUD

tus entre autres), *Baccharis spp. Trevoa trinervis, Proustia pungens*, quelques *Papilionaceaees* du genre *Adesmia*, en plus d'autres herbes et arbustes èpineux. Près du bord des fleuves, se développe un composante caractéristique: *Tessaria absinthioides*; et sur les roches, aux environs du littoral de La Serena, on trouve quelques *Nolanaceaes* qui se raréfient plus au sud. A partir de la montagne moyenne une steppe arbustive se développe, ouverte et basse avec une nette dominance de l'*Acacia caven*, accompagnée d'autres arbustes heliophytes comme *Colliguaya odorifera, Cestrum parqui, Muehlenbeckia hastulata, Proustia pungens*, etc. Un peu disperses on voit quelques arbres xerophytes (*Geoffroea decorticans*, Prosopis *chilensis* et *Porlieria chilensis*). Secondement, moins abondants, on trouve quelques arbres mesophytes qui ont une représentation majeure à partir de la vallée de l'Aconcagua, et qui remplacent les trois espèces antérieures. Il s'agit de *Cassia closiana, Lithraea caustica, Schinus latifolius* et *Quillaja saponaria.*

Il faut signaler que le bassin du fleuve Aconcagua est une barrière importante pour l'aire de distribution géographique de quelques espèces. Il constitue la limite méridionale de dispersion du *Prosopis chilensis* et du *Porlieria chilensis*, et du même c'est la limite septentrionale d'expansion d'arbres caduques tempérés comme *Nothofagus obliqua, Beilschmiedia miersii*, et *Crinodendron patagua*. Dans ce bassin il se trouve aussi la seule palme du territoire et de caractère tropicale: *Juabaea chilensis.*

Dans le bassin des fleuves Elqui et Limari, l'étage collinéen s'arrête entre les 400 et 500 mts. avec la disparition de *Acacia caven*. Du même ces terrasses fluviomarines de ces fleuves hébergent des espèces plus adaptées à l'écologie de montage comme c'est le cas du gros *Trichocereus, Prosopis chilensis* et *Porlieria chilensis*, en contact avec les arbres esclérophylles déjà cités. Cette végétation se rarifie avec la hauteur et aux environs de 1.800 mts. les espèces arborées disparaissent, et c'est *Prosopis chilensis* l'arbre qui a la plus grande dispersion altitudinale. On arrive ou étage subandin, avec un cortège floristique intéressant de petits arbustes et d'herbes héliophytes près de graminées dures des genres *Stipa* et *Festuca* lesquelles en endroits de majeur humidité cèdent la place à de petites prairies andines, et qui sont utilisées pour des activités de pâturage pendant l'été.

Proche au 3.800 et 4.000 mts. se trouve le milieu de l'étage andin avec des espèces spécifiques en forme de rosette comme *Baccharis tola* et *Laretia acaulis.* La ligne d'équillibre des langues glacières située par ici vers les 5.200 et 5.400 mts., signale la montée maximum de la végétation dans ces latitudes.

2. L'étagement au Chili Central *(33° au 38° Sud)* (Fig. 6, 7, 8, 9 et 10)

Nous incorporons ici la partie ou zone de tendance méditerranéenne atenuée, ou de type tempéré modérée. – Aux 38° Sud dans le bassin du fleuve Aconcagua, la dispersion altitudinale de la végétation se modifie un peu. La disparition des chaînes transversales du Norte Chico et l'effet modérateur du climat – surtout sur la cordillère côtière – lui donnent un plus vaste homogénéité et même une richesse relative au différents étages végétaux. Il faut remarquer toujours, que l'ac-

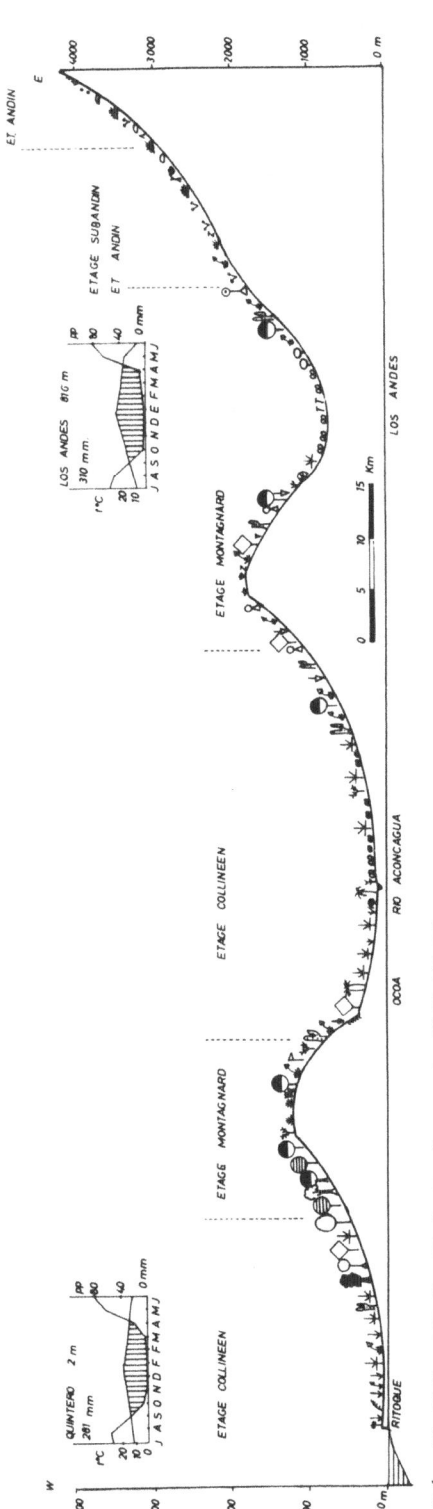

FIG. 6. — III. PROFIL PHYTOGEOGRAPHIQUE AUX ENVIRONS DU 32° 52' LAT SUD

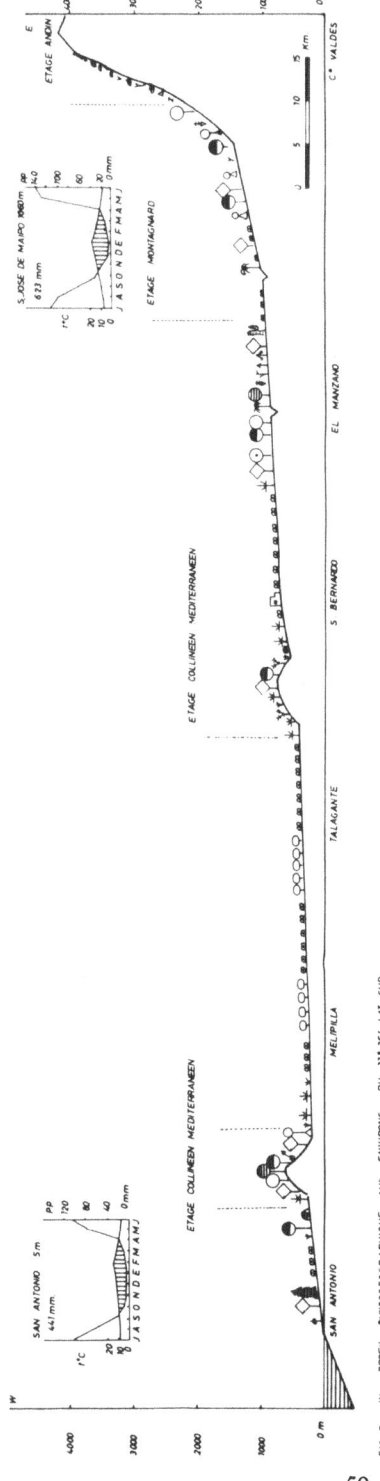

FIG. 7. — IV. PROFIL PHYTOGEOGRAPHIQUE AUX ENVIRONS DU 33° 36' LAT SUD

59

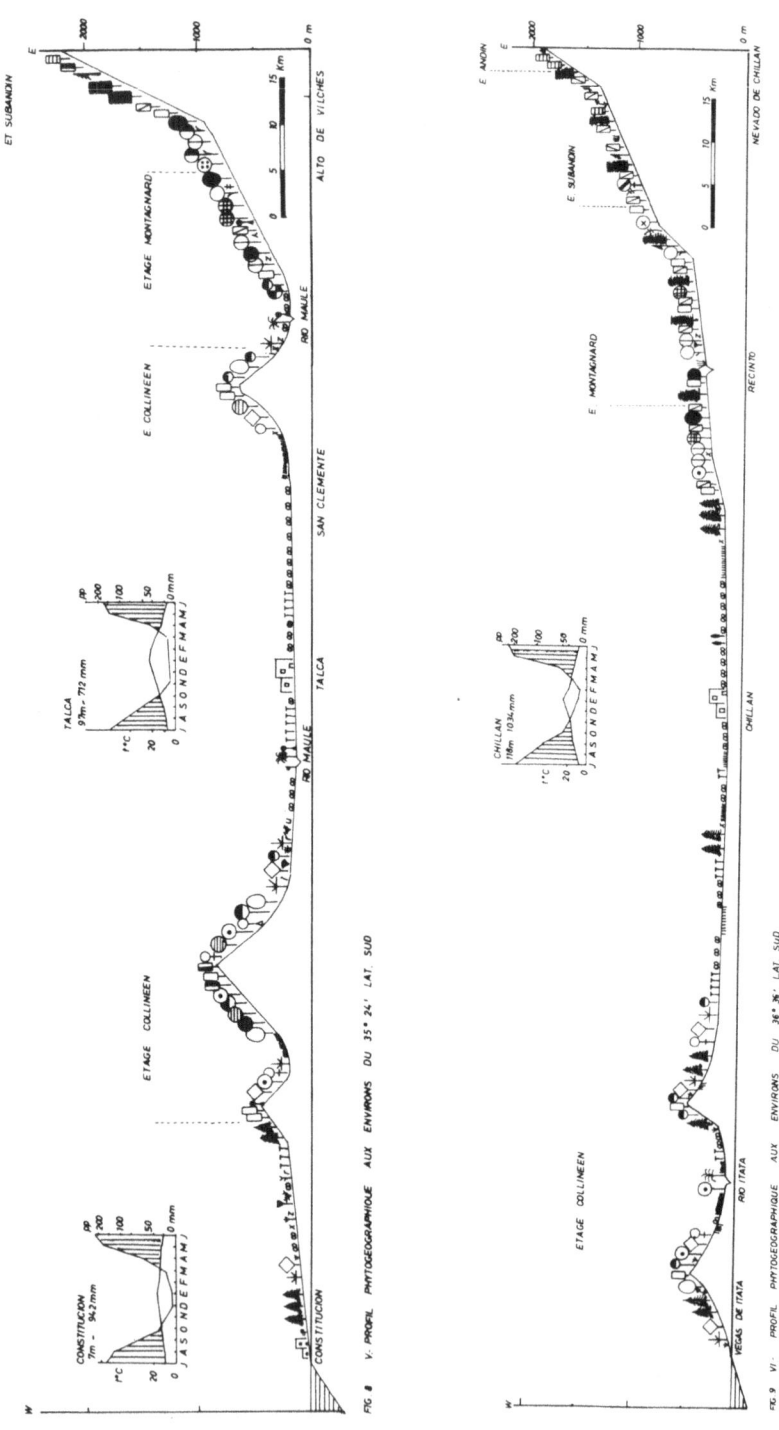

FIG. 8 V: PROFIL PHYTOGÉOGRAPHIQUE AUX ENVIRONS DU 35° 24′ LAT. SUD

FIG. 9 VI: PROFIL PHYTOGÉOGRAPHIQUE AUX ENVIRONS DU 36° 36′ LAT. SUD

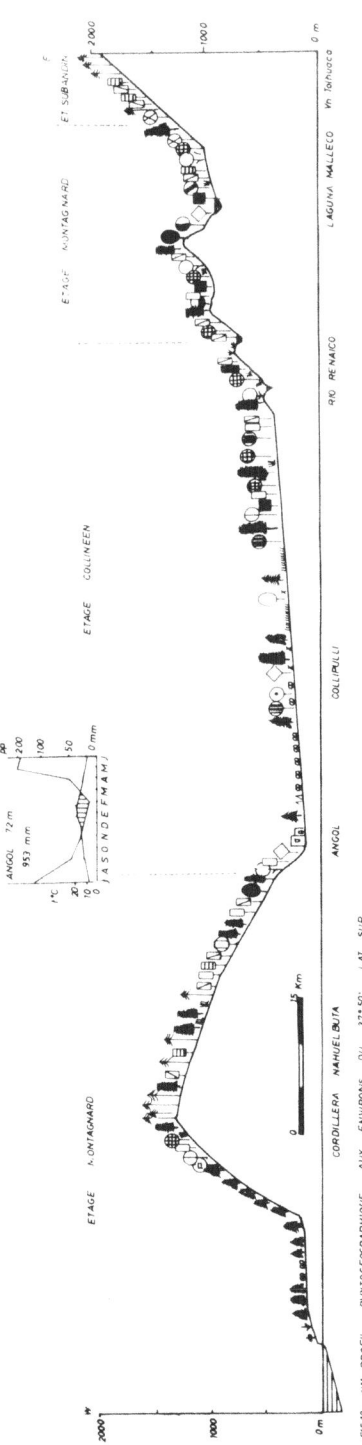

FIG.10.- VII PROFIL PHYTOGEOGRAPHIQUE AUX ENVIRONS DU 37° 50' LAT SUR

tion anthropogéne a altéré aussi de manière évidente les étages inférieurs.

A conséquence de l'avancement en latitude et des différents tendances qui présente le bioclimat méditerranéen au Chili tempéré (Quintanilla 1974); nous divisons cette zone en deux secteurs. Une aire de bioclimat de type méditerranéen qui s'étend a peu près entre les 33° et 35° et à laquelle correspondent les profils N° 7 et 8 et à continuation jusqu'aux 38° Sud, une aire de bioclimat méditerranéen atenué à laquelle correspondente les profils N° 9 et 10.

Les profils qui appartiennent au premier secteur, correspondent en grande partie aux bassins des fleuves Maipo et Maule.

Les deux secteurs sont caractérisés pour avoir un étage collinéen floristiquement similaire et alteré. Les collines de la Cordillère de la Côte dont les versants sont beaucoup plus exposés à l'influence océanique constituent l'habitat de la plupart des forêts sclérophylles, caracterisant de la même manière les endroits de l'étage montagnard mais en donnant lieu ou groupements arborés de quelque importance. La zone de contact entre les étages collinéen et montagnard, est principalement occupée par *Acacia caven* et d'autres espèces épineuses. Quand la forêt se développe degradée, l' 'espinal' et son cortège pénètre à l'intérieur en se mêlant en partie avec les arbres et en originant une fourrée connue avec le nom de 'matorral'.

Les sommets les plus hauts de la cordillère de la Côte dans cette région (massifs le Roble 2222 mts. et La Campana 1939 mts.) ont dès environ les 1400 mts. une végétation azonal, principalement à cause des avantages que lui donnent l'exposition océanique de ses versants. Communautés de *Nothofagus obliqua* avec des espèces higrophytes comme *Myrceugenia obtusa*, *Drimys winteri*, *Aristotelia chilensis* et *Chusquea cumingii* parmi d'autres, forment un 'monte' arboré et plus ou moins fermé, qui se ressemble beaucoup en physionomie et composition aux forêts tempérées du Chili austral.

L'étage montagnard se montre quand quelques espèces xerophytes (*Litrahea caustica*, *Trevoa trinervis*, etc.) disparaissent, et se déplace à peu près entre les 800 et 1800 mts. Aux environs des 1400-1500 mts. on distingue un subhorizon montagnard avec des arbres plus spécialisés aux effets des hauteurs. Il s'agit de *Azara petiolaris*, *Quillaja saponaria*, *Schinus montanus* qui a remplacé à *Schinus latifolius*, et *Kageneckia angustifolia* que d'autre part remplace a *Kageneckia oblonga*. Les *Kageneckias*, surtout *K. angustifolia*, représentent le maximum ascensionnel en hauteur des arbres sclérophylles. *Quillaja saponaria* monte jusqu'aux 1.800 mts., tandis qu'on trouve *Kageneckia angustifolia* même par dessus les 2000 mts.

Dans ces latitudes, l'étage subandin est très peu représentatif. A l'évidente faiblesse d'arbres sclérophylles, s'ajoute l'absence absolue des espèces résineuses. Ce caractère, plus l'existence de graminées dures, la forme entortillée et reduite de quelques arbustes demiépineuses (*Chuquiraga oppositifolia*, *Berberis empetrifolia*); signalent que nous montons sur un secteur de transition vers le milieu haut andin.

Autour des 2.500-2.600 mts. apparaissent des plantes qui restent sous la neige une grande partie de l'année: *Ephedra andina*, *Laretia acaulis* avec des graminées du genre *Festuca* et *Stipa*. Espèces en forme de coussines et disperses avec d'autre de beau développement estival (*Adesmia sp;*, *Viola sp.*), on les trouve jusqu'aux 3.500 mts. dans la limite entre les étages andin et nival. La ligne d'équilibre des glaciers, dans le massif d'Aconcagua se trouve environ les 4.000 mts., tandis que dans la cordillère de Santiago elle est vers les 3.700 mts. (Lliboutry, 1956).

Entre les 36° et 38° de latitude sud, les formations végétales des cordillères chiliennes sont plus riches dans le nombre d'espèces et dans la densité grâce à la majeur himidité et à la diminution de la période sèche tout le long de l'année. Malgré cela, le caractère du climat méditerréenne est encore présent dans les vallées et les versants orientaux de la cordillère de la côte ce qui se voit dans les hautes températures et dans la forte sécheresse du période estival.

Dans les secteurs moins modifiés par l'action anthropogéne, l'étage collinéen présente une dominance de la formation de l' 'espino', accompagnée de quelques arbres sclérophylles notamment *Lithraea caustica*, *Quillaja saponaria*, *Schinus polygamus* plus *Baccharis* spp.

L'étage montagnard des trois derniers profils (Fig. 8, 9 et 10) montre une structure physionomique et une composition floristique à peu près communes. D'abord il y a un subhorizon montagnard inférieur d'arbres sclérophylles lesquelles, sont au sud des 37° représentées presque exclusivement par *Peumus boldus*, *Maytenus boaria* et les *Lauraceaes Cryptocaria alba* et *Persea lingue*. Dans les endroits moins humides et d'une manière isolée existent *Quillaja saponaria* et *Lithraea caustica*. Des *Protacees* aussi comme *Gevuina avellana*, *Lomatia hirsuta* et *Embothrium coccineum*. Egalement une *Eleocarpacee*, *Aristotelia chilensis* et une *Winteraceae* typique des endroits humides *Drimys winteri*.

Autour des 700 mts. apparaît l'horizon montagnard supérieur, dénotant une dominance de Nothofagus notamment *Nothofagus obliqua*.

Dans la latitude du fleuve Maule (35° Sud) et dans la cordillère de Nahuelbuta (38° Sud), l'étage montagnard présente une particularité unique de la cordillère de la côte chilienne.

Dans les 35°30' S. l'horizon montagnard inférieur héberge des forêts très précieuses de trois types d'hêtres exclusifs de cette région du Chili. Il s'agit de *Nothofagus alessandri* Espinosa (le ruil), *Nothofagus leoni* Espinosa (roble maulino ou roble colorado) et *Nothofagus glauca* (Phil.) Krasser (roble blanco o hualo). Les deux derniers sont aussi représentés dans la cordillère des Andes et le *Nothofagus glauca* monte jusqu'au 1.000 mts. Dans la chaîne côtière, les trois *Fagaceaes* apparaissent vers les 350 mts. et on les trouve jusqu'au les 700 mts. a peu près. Ils sont souvent associés avec d'autres hêtres, abondants aussi dans les Andes, comme *Nothofagus obliqua* dans les parties basses et *Nothofagus alpina* dans les parties supérieures. Les hêtres du bassin du Maule ont eu dans le passé une distribution géographique plus importante sur la cordillère de la Côte, mais le main de l'homme bûcheron, constructeur et charbonnier finit par le réduire seulement à quelques centaines d'hectares et qu'auhourd'hui se protègent grâce au régime de forêt domaniale.

Dans la cordillère de Nahuelbuta, entre les 37° et 38° la chaîne côtière dépasse les 1.300 mts. et l'étage montagnard supérieur présente une autre intéressante curiosité phytogéographique. Il s'agit de l'existence d'*Araucaria araucana* K. conifère endémique de l'Amérique du Sud qui est ici représentée de préférence dans la partie septentrionale de cette cordillère locale. Elle apparaît aux 700 mts. associée avec *Nothofagus dombeyi* et *Nothofagus pumilio* (Schulmeyer 1978). Dans la versant pacifique côtière, ce secteur correspond à l'unique station écologique qui possède cette resineuse qui a une repartition plus vaste dans la cordillère andine et qui monte jusqu'aux 2.000 mts. presque.

Dans la cordillère des Andes l'étage montagnard se modifie un peu en direction le sud, principalement comme conséquence de l'augmentation de l'humidité. La forêt caduque de *Nothofagus*, dominant dans les cours supérieurs des bassins hydrographiques des fleuves Maule et Itata (Profil V et VI), est remplacée vers les 38° S. par la forêt pluvieuse sempervirente (Profil VII). L'étage montagnard inférieur montre encore des espèces sclérophylles comme *Cryptocaria alba* et *Peumus boldus*. Après, le bois se serre en plus avec la présence de quelques *Lauraceaes* et *Protaceaes* et *Fagaceaes* comme *Nothofagus obliqua*, *Nothofagus alpina* et *Nothofagus dombeyi*. Cette forêt possède un strate inférieur plus au moins dense à l'intérieur duquel on se détachent *Chusquea quila*, *Aristotelia chilensis* et de grandes herbes.

L'apparition des premières conifères correspondantes à la famille *Podocarpaceaes* vient à signaler le terme du milieu montagnard et la présence de l'étage subandin. A exception du cas de la cordillère andine de Talca, reste seulement trois Nothofagus sur les 1.400 mts. Ils sont: *Nothofagus dombeyi*, *Nothofagus pumilio* et *Nothofagus antarctica*, et c'est le dernier, l'arbre caduque celui qui montre le plus considérable dispersion altitudinale et même il constitue, dans cette région, la limite altitudinale supérieure des arbres entre les 2.000 et 2.200 mts. L'*Araucaria araucana* seulement, entre les 37° et 40° sud, dépasse cet avancement en hauteur.

En général, les étages andins de la cordillère chiliienne homonyme, montrent un relative appauvrissement d'espèces spécifiques par rapport à la richesse des étages inférieurs. Uniquement les forêts de *Nothofagus alpina* et *Nothofagus antarctica* en outre des conifères (*Podocarpus spp.*, *Austrocedrus chilensis*) sont des groupements typiques des Andes. Quelques espèces d'étages inférieurs comme les *Lauraceaes*, *Drimys winteri* et les arbustes humides, ont disparu. Le soubois est ouvert et pauvre.

En résumé, l'étage andin, comme le signale Oberdorfer (1960), entre les 1.800 et les 2.200 mts. est composé par des associations méridionales appauvries de caducifoliés subantarctiques comme *Nothofagus pumilio* et *Nothofagus antarctica*; ou plus bas, par des associations mixtes de *Podocarpus andinus* et les caduques. Ces groupements, laissent dans la périphérie de la forêt des endroits avec un peu de afleurement d'eau ou, en plus de petits prairies andines, on peut trouver dans ces altitudes de petites taches de *Chusquea quila*, particulièrement sur les cordillères de Talca et Chillàn. D'autre part sur les versants du volcan Tolhuaca, *Araucaria araucana* conforme des associations pures en constituant a la fois le dernier ceinturon arboré qui monte sur les Andes, dans cette latitude.

Conclusions

Dans ce bref coup d'oeil sur la dispertion altitudinale de la végétation dans les cordillères du Chili Central, nous constatons à notre avis, deux faits importants.

D'abord, la difficulté de clarifier les étages végétaux avec des niveaux altimétriques et des formations physionomiques fixes. Cela obéit à des diverses raisons: à la vaste amplitude écologique altitudinale qui ont quelques espèces; c'est le cas de *Quillaja saponaria* et les *Kageneckias* parmi les arbres sclérophylles, et de *Nothofagus alpina* et *Nothofagus dombeyi* parmi les *Fagaceaes*. Ensuite, la diminution de la hauteur absolue des cordillères parallèle avec l'avancement de la latitude vers le sud, accompagnée avec un augmentation des conditions de l'humidité; rend difficile de trouver une homogénité et une amplitude précise pour les étages, et même en tenant compte de leur diversité floristique. L'intervention anthropogéne d'autre part, très avancée sur les cordillères du Norte Chico et ligère encore sur celles du sud compliquent aussi l'identification des paysages autochtones. On trouve des espèces introduites près des 2.000 mts. (Exemple: *Pinus radiata*).

Une deuxième consideration pour nous, c'est qu'au niveau local et régional l'analyse descriptif de la distribution altitudinale de la végétation, permet aujourd'hui mieux carcatériser la géobotanique des régions biogéographiques chiliennes. Dans un gradient altimétrique ont peut étudier plus en détail les effets des facteurs locaux, notamment pour le cas du relief et du climat, et arriver à mieux expliquer les biomas régionaux des cordillères. Dans un pays de montagnes comme c'est le Chili, la végétation se subordonne aussi aux influences orographiques. L'utilisation de profils altitudinaux phytogéographiques, est une technique intéressante pour aider aux études de la végétation.

Remerciements: Nous devons exprimer nos remerciements à Mr. Garaventa et aux Professeurs O. Zollnner de la Université Catholique de Valparaíso, et P. Aravena de la Université Catholique du Chili (Talca); pour leur appui botanique à nos recherches.

Resumen

Por su larga extensión latitudinal, Chile posee varios tipos de climas, lo cual incide a su vez en la diversidad de su vegetación. Nuestra zona de trabajo se sitúa entre los 30° y 38° Sur y de modo general se encuentra al interior de los climas de tendencia mediterrànea. Éstos, a medida que se avanza hacia las latitudes meridionales, van moderando sus caracteres especialmente en lo que respecta a la duración del período seco. El territorio chileno, con un ancho promedio de 110 kms., està flanqueado por dos largas cadenas de montañas como son la Cordillera de la Costa al oeste y la Cordillera de los Andes al este, y ellas influencian notoriamente los tipos humanos de vida y la explotación de los recursos naturales del Valle Longitudinal; como también al clima y a la vegetación.

Desde el punto de vista biogeogràfico regional, hasta los 33° Sur, Chile posee una vegetación esteparia de tipo semiàrido, con arbustos espinosos y especies suculentas. A partir de la cuenca del río Aconcagua (33° S.), el mejoramiento gradual de las condiciones ecológicas permiten el surgimiento del bosque esclerófilo, preferentemente cantonado en las montañas, y de la estepa semixerófita de Acacia caven en los sectores bajos y abiertos de Chile interior. Solo los sectores hùmedos de las montañas, cobijan especies templadas. A continuación de los 35° S., el aumento de la humedad y la mejor calidad de los suelos, otorgan un nuevo paisaje a la vegetación notablemente caracterizada por los bosques de *Nothofagus* y *Lauràceas*. Desde los 37°, las agrupaciones vegetales se enriquecen aùn màs, con la incorporación de especies higrófitas y sobre todo de las coníferas chilenas. Particularmente interesantes son los bosques de *Araucaria araucana*, los cuales se distribuyen en ambas cordilleras aunque en mayor extensión y altura en los Andes.

Actualmente, son precisamente las cordilleras quienes revisten una gran importancia fitogeogràfica dado que los efectos antrópicos y geofísicos (en especial los climàticos), han raleado y deteriorado enormemente los paisajes vegetales de las depresiones centrales y valles fluviales, especialmente entre los 30° y 36° Sur.

El criterio de diferenciación de pisos vegetales es muy ùtil para clasificar las formaciones vegetales que se suceden en altura. Algunas de ellas, a causa del avance en latitudud hacia el sur y del descenso de la cordillera andina en la misma dirección, poseen una distribución geogràfica màs o menos extensa. Dentro del àrea de clima mediterràneo de tipo acentuado, es la cordillera de la costa — a causa de recibir el mayor efecto de la influencia oceànica — quien posee ùnicamente formaciones vegetales àrboreas como los bosques esclerófilos, desarrollàndose ademàs en algunos macizos, pequeñas agrupaciones de caràcter templado sobre todo a base de *Nothofagus*. Los Andes en estas latitudes, poseen grupos dispersos de àboles esclerófilos y especies espinosas.

Al sur de los 35°, la vegetación nativa recubre gran parte de las vertientes de las dos cordilleras. A esta latitud, la Cordillera de la Costa posee bosques monoespecíficos de *Nothofagus glauca* y *Nothofagus alessandri*; en tanto que en los Andes aparecen los *Podocarpus* como las primeras coníferas chilenas.

El piso Codineano o Basal sólo posee como dominantes a *Cryptocaria alba* y *Peumus boldus* entre los àrboles esclerófilos, en tanto que la predominancia de las *Fagàceas, Protàceas* y *Lauràceas* es evidente. Destacan *Nothofagus obliqua* y *Nothofagus alpina* en los Andes y también en la cordillera costera, pero a partir de los 36°. Por sobre los 700 u 800 metros surge el piso Montañoso en el cual, a los dos *Nothofagus* anteriores, se agrega *Nothofagus dombeyi* quien remonta hasta alrededor de los 1600 metros, altitud en la cual al hacerce presentes los *Podocarpus*, dan paso al piso Subandino.

En la costa, las alturas superiores de sus cadenas no sobrepasan los 1.400 metros en la Cordillera de Nahuelbuta, cuyas cimas estàn coronadas por la formación *Araucaria araucana.*

En los Andes, bosques abiertos de *Nothofagus pumilio* y *Nothofagus antarcti-*

ca, entre los 2000-2200 metros, indican el àmbito del piso andino donde por sobre ellos, arbustos bajos y leñosos junto con gramíneas duras compiten en espacio con las nieves eternas y lenguas glaciares. Sólo al sur de los 37° S y alrededor de los 2000 metros, crece el bosque de *Araucaria araucana* en la cordillera andina.

Puede decirse que màs o menos desde los 36°, tanto en la cadena costera como en los Andes, se denota una sucesión màs o menos clara de pisos vegetales. En la costa, las reforestaciones de *Pinus insignis*, resinosa norteamericana introducida, han alterado mucho el paisaje vegetal nativo del piso colineano y parte del montañoso. En los Andes este fenómeno, por ahora se presenta de modo disperso.

Bibliographie

Antonioletti, R., J. Borcosque, H. Schneider & E. Zàrate. 1971. Características climáticas del Norte Chico (26° a 33° lat. S). Inst. Inv. de Rec. Nat. Santiago.

Aravena, P. 1976. Los robles-Nothofagus de la Séptima Región de Chile, Región del Maule. *Rev. Maule*, Univ. Cat. de Chile, N° 3: 3-8.

di Castri, F. 1968. Esquisse écologique du Chili. Biologie de l'Amérique Australe; Editions C.N.R.S., vol. IV:, Paris.

di Castri, F. & Hajek, E. 1976. Bioclimatología de Chile. Vicerrec Acad. Univ. Cat. de Chile, Santiago.

Fuenzalida, V.H. 1965. Biogeografía de Chile. Geografía Económica de Chile. Texto Ref.: 228-267. Santiago.

Garaventa, A. 1936. Por qué componentes vegetales estarían constituidos los climax en la provincia de Aconcagua. Com. au IX Congres. Cient.: 54-64, Santiago.

Hajek, E., M. Pacheco & A. Passalacqua 1972. Anàlisis bioclimàtico de la sequía en la zona de tendencia mediterránea de Chile. Pub. N° 45 Inst. Geogr. Univ. Cat. de Chile, Santiago.

Hajek, E. F. di Castri 1975. Bioclimatografía de Chile. Dir. Invest. Univ. Cat. de Chile, Santiago.

Huber, A. 1977. Aporte a la climatología y climaecologìa de Chile I: Radiación potencial. *Rev. Medio Ambiente*, 2(2): 22-34. Valdivia.

Hueck, K. 1966. Die Wälder Südamerikas. Stuttgart G. Fisher.

Klapp, E. 1956. Futterbau und Futterwirtschaft in Chile zwischen dem 30° und 42° S. *Bonner Geograph*. Abh. 17: 87-136.

Lauer, W. 1952. Humide und aride Jahreszeiten in Afrika und Südamerika und ihre Beziehungen zu den Vegetationsgürteln. *Bonner Geograph*. Abh. 9: 1-98.

Looser, G. 1929. Diferencias entre la vegetación de la cordillera de la Costa y la cordillera de los Andes en Chile Central. *Rev. Univ. de Chile* Año XIV, Santiago.

Liiboutry, L. 1956. Nieves y glaciares de Chile. Ed. Univ. de Chile.

Mann, G. 1960. Regiones biogeográficas de Chile. *Rev. Inv. Zool. chilenas* 6: 15-49, Santiago.

Muñoz, C. & E. Pisano 1947. Estudio de la vegetación y flora de los Parques nacionales de Fray Jorge y Talinay. *Agri. Téc.* Año VII N° 2: 71-199, Santiago.

Muñoz, C. 1966. Sinopsis de la Flora chilena. Ed. Universidad de Chile, Santiago.

Oberdorfer, E. 1960. Pflanzensoziologische Studien in Chile ein Vergleich mit Europa. Flora et vegetatio mundi, Band II. J. Cramer Weinheim.

Ozenda, P. 1960-61. La détermination de la zone montagneuse à l'aide des limites altitudinales de végétation. Bull. de la Fed. française d'Eco. Montagnarde; N° 11, 4: 50-57.

Pisano, E. 1956. Esquema de clasificación de las comunidades vegetales de Chile. *Agronomía II*, N° 1, Santiago.

Pisano, E. 1966. Zonas biogeográficas de Chile. Geogr. Econom. de Chile, Primer Apend.: 62-72, Santiago.

Quintanilla, V. 1974. Les formations végétales du Chili Tempéré. Essai écologique et phyto-géographique. *Doc. de Cartograph. Ecolog.*: 38-80, vol. XIV. Grenoble.

Quintanilla, V. 1974. La Carta bioclimática de Chile Central. *Rev. Geogr. Valparaíso* N° 5: 33-58, Valparaíso.

Quintanilla, V. 1977. A contribution to the phytogeographical study of temperate Chile. *Biogeographica* 8: 31-41, The Hague.

Quintanilla, V. 1977. Diccionario de Biogeografía para América Latina. Ed. Universitarias de Valparaíso, 250 p. Chile.

Quintanilla, V. 1978. Los perfiles fitogeográficos del medio seminàrido de Chile (Norte Chico). *Rev. Geogr.* Inst. Panam. Hist. y Geogr. N° 88 (sous presse).

Reiche, C. 1934. Geografía Botánica de Chile. 2 vols. Trad. de G. Looser. Imprenta Universitaria, Santiago.

Schmithüsen, J. 1954. Waldgesellschaften des Nördlichen Mittelchile. *Vegetatio* vol. V-VI: 479-486, The Hague.

Schmithüsen, J. 1956. Die raümliche Ordnung des chilenischen Vegetation. *Bonner Geogr.* Abh. 17: 1-89, Bonn.

Schneider, H. 1968. El clima del Norte Chico. Depto. de Geogr. Univ. de Chile, Santiago.

Schulmeyer, D. 1978. Observaciones fitogeográficas sobre la Cordillera de Nahuelbuta. Bol. Informativo. II trimest., Inst. Geogr. Militar, Santiago.

Villagràn, C. 1969. Notas palinológicas de los bosques relictuales de la zona Central de Chile. *Notic Mens. Museo Hist. Nat.* N° 153: 3-6, Santiago.

Adress of the Author:
Prof. Dr. Victor G. Quintanilla,
Facultad de Ingienería
Departamento de Geodesia
Universidad Técnica del Estado Avda
Ecuador 3469
Santiago
CHILE.

GEBÜSCHGESELLSCHAFTEN IM MITTELEUROPÄISCHEN LANDSCHAFTSBILD

M. Moor

Als Endglied ungestörter Vegetationsentwicklung, als Klimax also, dürften in der planaren und kollinen Stufe des gemässigten Mitteleuropas die Eichen-Birken- und die Eichen-Hagebuchenwälder betrachtet werden, die bei einigermassen gleichbleibenden grossklimatischen Verhältnissen das mit Boden, Klima und Lebewesen ausbalancierte Gleichgewicht darstellen.

Liegen jedoch nicht mittlere Verhältnisse vor, sind vielmehr unter den edaphischen und den klimatischen Faktoren oder unter den biotischen Faktoren solche, die die Entwicklung aufhalten oder in eine andere Richtung lenken, dann stellen bestimmte Glieder der Entwicklungsreihen an den betreffenden Standorten einen vorläufigen Endzustand dar, sog. Dauergesellschaften, die so lange den betreffenden Standort besetzen, als der besondere Faktor wirkt. Zu den besonderen Faktoren zu zählen sind auch die menschlichen Eingriffe ins Pflanzenkleid wie Rodung, Mahd, Tritt oder Beweidung. Zu den Dauergesellschaften zählen nicht nur Krautfluren und Rasen, sondern auch bestimmte Wälder, so die Schwarzerlenbrücher auf dauernd vernässten Torfböden, der Silberweidenwald auf periodisch überschwemmten sandigen Terrassen oder die Föhrenwälder der Mergelrutschhalden und der Felsgräte, um nur einige wenige Beispiele zu nennen – Spezialisten oder Sondergesellschaften auf ebensolchen Spezial- oder Sonderstandorten.

Gebüschgesellschaften sind im gemässigten Mitteleuropa nur dort zu erwarten, wo der Wald – gleichgültig ob Klimax- oder Dauergesellschaft – an waldfeindliches Gelände grenzt. Nur solchen Grenzlinien entlang vermögen Straucharten zu Gesellschaften zusammenzuschliessen. Es sind die sog. Mantelgesellschaften.

In unteren Lagen Mitteleuropas, in der Ebene und in der Hügelstufe, sind waldfeindliche Standorte die Ausnahme. Dazu zählen z.B. die Dünen am Meeresstrand, die Felswände und Felsschutthalden und die offenen Wasserflächen der stehenden und fliessenden Gewässer. Aber auch die Rodungsflächen, die der Mensch bebaut und auf denen sich seine Siedlungen ausdehnen, stellen Standorte dar, wo Beweidung und Tritt, Mahd und erst recht Hacken und Pflügen den Wald am Aufkommen verhindern.

Infolge der Rodungen haben Waldränder riesenhafte Ausdehnung erlangt, und Gebüschgesellschaften haben ungewollt reiche Entfaltungsmöglichkeiten erhalten. Es sind aber nur halbnatürliche Gesellschaften, die in den letzten zwei oder drei Jahrtausenden auf gerodetem Waldboden neu entstanden sind. Natürliche Gebüschgesellschaften sind im Waldland Mitteleuropa die Ausnahme.

Ein augenfälliges Merkmal dieser Gebüschgesellschaften ist der Standort, entfalten sie sich doch, wie oben schon erwähnt, dort, wo der Wald an baum- oder gar vegetationsloses Gelände grenzt. Der Standort der Gebüschgesellschaften zwischen Wald und Krautflur ergibt eine ästhetisch hoch befriedigende Überleitung von baumhafter Vegetation zu Krautfluren oder zu völliger 'Wüste'. Offene Wasserflächen, nackter Fels und nackte Felsschutthalden, vom Wind bewegte Sanddünen so gut wie menschliche Siedlungsflächen müssen, von der Vegetation aus gesehen,

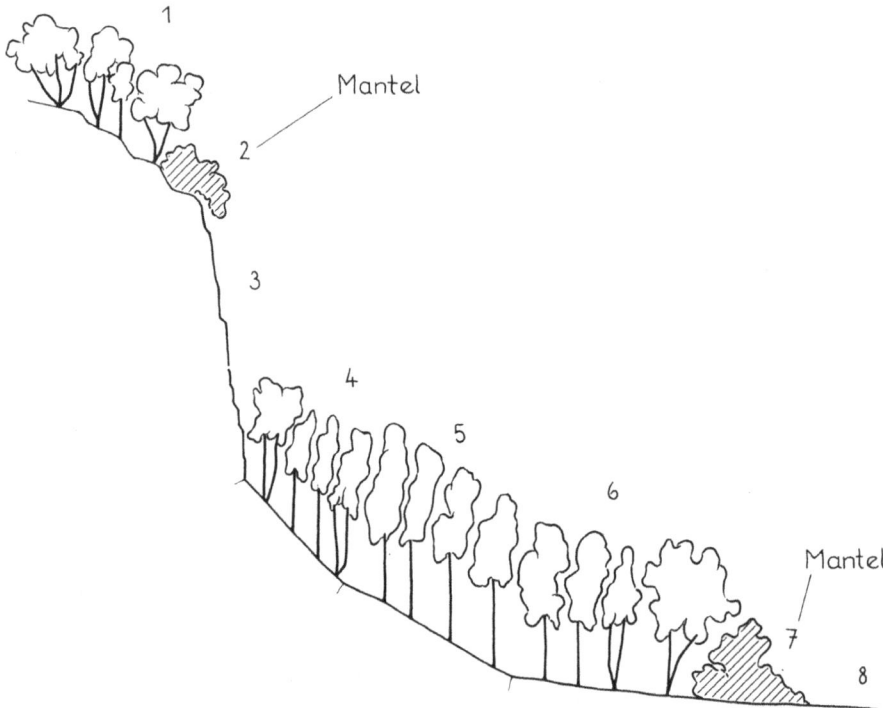

Fig. 1. Mantelgesellschaften im Kalkgebiet des Schweizer Juras

1 Flaumeichenbuschwald
2 *Felsenbirnengebüsch*
3 Felsspaltengesellschaft
4 Lindenmischwald
5 Seggen-Buchenwald
6 Eichen-Hagebuchenwald
7 *Liguster-Schlehengebüsch*
8 Halbtrockenrasen

Felstreppen und Mähwiese sind baumfreie Standorte, der eine von Natur aus, der andere künstlich.
Flaumeichenbuschwald und Laubmischwald sind durch eine Gebüschgesellschaft gegen das waldfreie Gelände abgeschlossen.
Das Felsenbirnengebüsch ist eine natürliche Mantelgesellschaft, der Liguster-Schlehenbusch eine halbnatürliche.
(vgl. M. Moor 1960, 1962 und 1964)

doch wohl als 'Wüste' bezeichnet werden, als wüst und leer, weil vegetationslos.

Folgende Aufzählung stellt die im gemässigten Mitteleuropa vorkommenden waldfreien Standorte zusammen, nennt den jeweiligen waldfeindlichen Faktor und versucht, die Folge von Krautflur über Gebüschgesellschaft zu Wald herauszustellen. Dabei handelt es sich entweder um ein bloss räumliches Nebeneinander, um eine sog. Zonation, oder aber um ein zeitliches Nacheinander, wenn eine Sukzession vorliegt.

Die Reihenfolge der aufgezählten Dinge ist in allen 14 Fällen dieselbe, nämlich:

a. Waldfreier Standort, im Extrem vegetationslos

b. Baum- oder vegetationsfeindlicher Faktor

c. Zonation (und Sukzession) in der Pflanzendecke
1. Krautflur   2. Gebüsch   3. Wald.

Die Nomenklatur der Pflanzengesellschaften hält sich an E. Oberdorfer 1970 p. 22-41.

I. a. Tümpel, Weiher, See
    b. Dauernd hoher Wasserstand
    c. 1. Wasserlinsen-Gesellschaften (Lemnetea)
          Laichkraut- und Schwimmblatt-Gesellschaften (Potamogetonetea)
          Strandlings-Gesellschaften (Littorelletea)
          Röhricht- und Grosseggen-Gesellschaften (Phragmitetea)
       2. Faulbaum-Aschweide-Gebüsch (*Frangulo-Salicetum*)
          Oehrchenweide-Gebüsch (*Salicetum auritae*)
       3. Schwarzerlenbruchwald (*Carici elongatae-Alnetum glutinosae*)

II. a. Bach, Fluss, Strom
     b. Periodische Überflutung mit fliessendem Wasser
     c. 1. Röhrichte (Phragmitetea)
        2. Mandelweiden-Gebüsch (*Salicetum triandro-viminalis*)
        3. Silberweidenwald (*Salicetum albo-fragilis*)

III. a. Quellnischen, Quelltöpfe, Tuffhügel
     b. Dauernde Überrieselung mit Quellwasser
     c. 1. Quellfluren (*Montio-Cardaminetea*)
        2. Pfaffenhütchen-Schwarzholunder-Gebüsch (*Evonymo-Sambucetum*)
        3. Ahorn-Eschenwald (*Aceri-Fraxinetum*)
           Bacheschenwald (*Carici remotae – Fraxinetum*)

IV. a. Ufer von Seen und Flüssen
    b. Grosse Wasserstandsschwankungen, Wechsel von Überflutung und Austrocknung
    c. 1. Zwergbinsengesellschaften (Nanojuncetea)
          Schlammufer-Gesellschaften (Bidentetea)
          Flutrasen (Plantaginetea)
       2. Schwarzweide-Schneeball-Gebüsch (*Salici-Viburnetum*)
          Alpenschwarzweide-Gebüsch (*Salicetum alpicolae*)
       3. Schachtelhalm-Grauerlenwald (*Equiseto-Alnetum incanae*)
          Reitgras-Grauerlenwald (*Calamagrosti-Alnetum incanae*)

V. a. Kiesbuckel, Kiesflächen und Sandbänke in Flussbetten und in der Aue
    b. Frische Aufschüttungen, Überführung mit Sand und Kies durch fliessendes Wasser
    c. 1. Geröllfluren (Thlaspietea)
          Ruderale Ufersäume (Artemisietea)
       2. Sanddorn-Sauerdorn-Gebüsch (*Hippophao-Berberidetum*)
       3. Wintergrün-Föhrenwald (*Pyrolo-Pinetum*)

VI.   a. Dünen der Meeresküste und des Binnenlandes
       b. Ständige Verfrachtung (Instabilität) des Bodens durch den Wind
       c. 1. Strandhafer-Dünen (Ammophiletea)
           2. Sanddorn-Liguster-Gebüsch (*Hippophao-Ligustretum*)
           3. Eichen-Birkenwald (*Querco-Betuletum*)
VII. a. Felsköpfe, Felsgräte und treppige Felsabstürze (anstehender Fels)
       b. Sehr kleiner Wurzelraum, windexponiert und stark austrocknend
       c. 1. Felsspalten-Gesellschaften (Asplenietea)
              Felsgrusfluren (*Sedo-Scleranthetea*)
              Trockenrasen, Magerrasen (*Festuco-Brometea*)
           2. Felsenbirnen-Gebüsch (*Cotoneastro-Amelanchieretum*)
           3. Flaumeichenwald (*Coronillo-Quercetum*)
              Blaugras-Buchenwald (*Seslerio-Fagetum*)
              Kronwicken-Föhrenwald (*Coronillo-Pinetum*)
              Bärlapp-Bergföhrenwald (*Lycopodio-Mugetum*)
VIII. a. Felsschutthalden, Schotter- und Nagelfluhrutschhänge
       b. Feinerdearmut im Oberboden und Instabilität der Bodenoberfläche
       c. 1. Steinschutt-Gesellschaften (Thlaspietea)
           2. Maiglöckchen-Haselgebüsch (*Convallario-Coryletum*)
              Weiden-Alpenkreuzdorn-Gebüsch (*Salici grandifoliae-Rhamnetum alpinae*)
           3. Hirschzungen-Ahornwald (*Phyllitido-Aceretum*)
              Ahorn-Lindenwald (*Aceri-Tilietum*)
              Mehlbeer-Ahornwald (*Sorbo-Aceretum*)

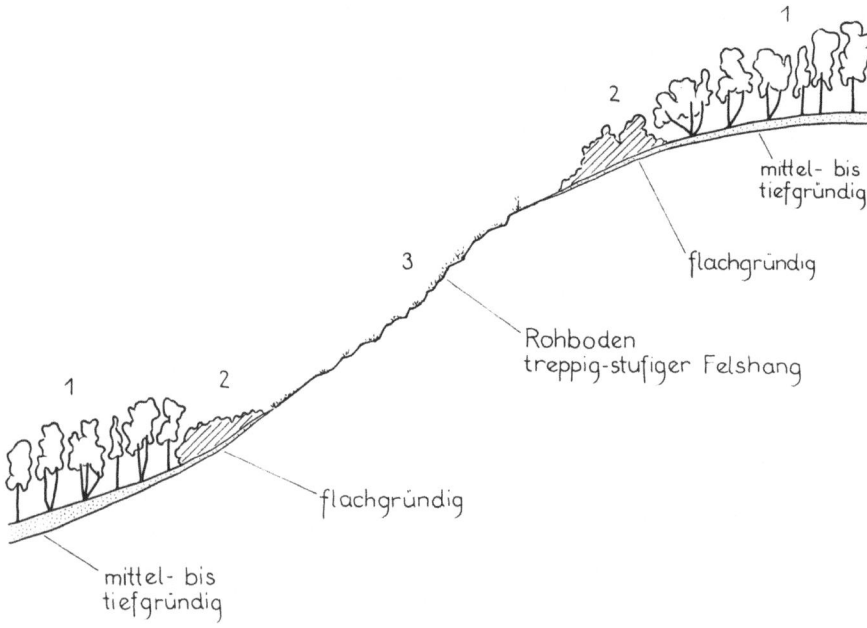

Fig. 2. Verteilung von Wald, Gebüsch und Rasen in den Kalkbergen Nordost-Serbiens

1 Buchenwald
2 *Syringetum*, eine natürliche Fliedervorholzgesellschaft zwischen Felsrasen und Wald
3 Felsrasen
(vgl. R. Tüxen 1952 p. 109, nach R. Knapp 1944)

IX. a. Rutschschneesteilhänge, Lawinenbahnen
    b. Abrutschender, auskämmender Schnee
    c. 1. Blaugras-Halden (*Elyno-Seslerietea*)
       2. Hasenohr-Mehlbeer-Gebüsch (*Bupleuro-Sorbetum*)
       3. Blaugras-Buchenwald (*Seslerio-Fagetum*)
X. a. Depotstellen von Lawinen- und Treibschnee
   b. Langandauernde Schneebedeckung
   c. 1. Subalpine Hochstaudenfluren (*Betulo-Adenostyletea*)
        Borstgras-Triften (*Nardo-Callunetea*)
      2. Ribes petraeum-Gesellschaft
        Lonicera coerulea-Rosa pendulina-Gesellschaft
      3. Hochstauden-Buchenwald (*Aceri-Fagetum*)
        Fichtenwälder (*Vaccinio-Piceetea*)
XI. a. Bülten und Schlenken in Hochmooren
    b. Torf, dauernd nass und nährstoffarm
    c. 1. Kleinseggen-Sümpfe (*Scheuchzerio-Cariceta*)
         Hochmoor-Bulte (*Oxycocco-Sphagnetea*)
       2. Oehrchenweide-Gebüsch (*Salicetum auritae*)
         Betula nana-Gebüsch
       3. Torfmoos-Bergföhrenwald (*Sphagno-Mugetum*)
XII. a. Salzwiesen der Meeresküsten und des Binnenlandes
     b. Salzgehalt des Bodens
     c. 1. Seegraswiesen (Zosteretea)
          Queller-Gesellschaften (*Thero-Salicornietea*)
          Salzschlickgraswiesen (Spartinetea)
          Salzwiesen (*Juncetea maritimae*)
        2. Sandweide-Gebüsch (*Salicetum arenariae*)
        3. Bodensaure Eichen-Birkenwälder (*Quercetea robori-petraeae*)
XIII. a. Kahlschlagflächen im Waldgebiet
      b. Blosstellung des Waldbodens
      c. 1. Schlagfluren (*Epilobietea angustifolii*)
         2. Weidenröschen-Salweide-Gebüsch (*Epilobio-Salicetum capreae*)
           Traubenholunder-Gebüsch (*Sambucetum racemosae*)
         3. Edellaubholzwälder (*Fraxino-Fagetea*)
XIV. a. Gerodetes, bebautes Gebiet
       (Aecker, Gartenland, Rebgelände, Weiden und Trittstellen, Nasswiesen, Frischwie-
       sen, Halbtrocken- und Trockenrasen)
     b. Menschlicher Eingriff
       (Roden, Hacken, Pflügen, Weidegang, Tritt, Mahd)
     c. 1. Hackunkraut-Gesellschaften (Chenopodietea)
          Getreideunkraut-Gesellschaften (Secalinetea)
          Trittrasen (Plantaginetea)
          Grünland-Gesellschaften [Fettwiesen, Fettweiden, Nasswiesen, Pfeifengras-Ried-
            wiesen] (*Molinio- Arrhenatheretea*)
          Trockenrasen, Magerrasen (*Festuco-Brometea*)
        2. Traubenkirschen-Haselgebüsch (*Pado-Coryletum*)
          Schwarzweide-Schneeball-Gebüsch (*Salici-Viburnetum*)
          Liguster-Schlehengebüsch (Ligustro-Prunetum)
        3. Eichen-Hagebuchenwälder (Carpinion)
          Buchenwälder (Fagion)
          Grauerlen- und Eschenwälder (*Alno-Padion*)
          Flaumeichenwälder (*Quercion pubescenti-petraeae*)

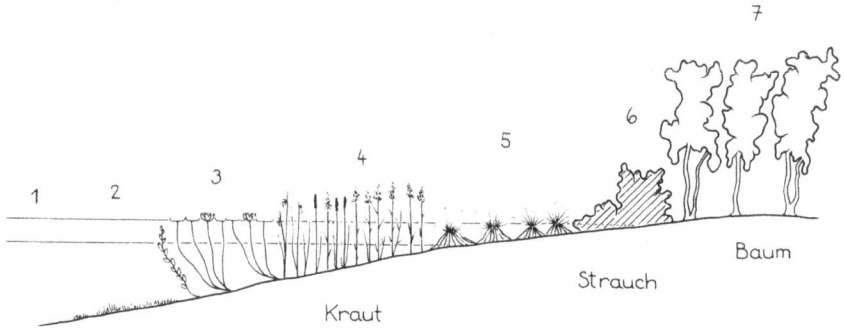

Fig. 3. Zonation der Pflanzengesellschaften am Ufer eines stehenden Gewässers (Mitteleuropa)

1 Offene Wasserfläche   vegetationslos
2 Armleuchteralgen-Wiese, submers
3 Laichkraut-Seerosen-Gesellschaft   } Krautfluren
4 Schilf-Binsen-Röhricht
5 Steifseggen-Wiese
6 *Oehrchenweide-Faulbaum-Gebüsch*   *Gebüschgesellschaft*
7 Schwarzerlen-Bruchwald   Wald

In gesetzmässiger Weise folgen sich bei abnehmender Wassertiefe Krautflur, Gebüsch und Wald.

Das räumliche Nebeneinander der verschiedenen Pflanzengesellschaften entspricht am Ufer eines stehenden Gewässers auch der genetischen Sukzession bei der Verlandung.

(vgl. M. Moor 1962 und 1969)

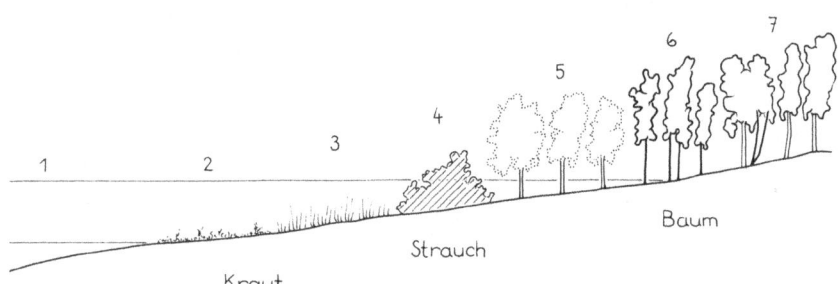

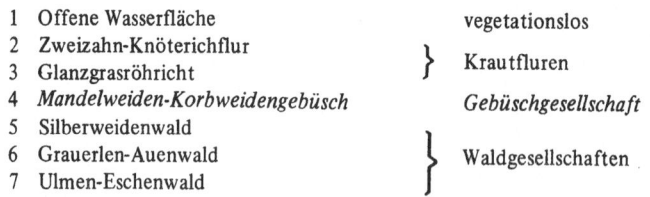

Fig. 4. Zonation der Pflanzengesellschaften in den Flussauen des Alpenvorlandes

1 Offene Wasserfläche   vegetationslos
2 Zweizahn-Knöterichflur
3 Glanzgrasröhricht   } Krautfluren
4 *Mandelweiden-Korbweidengebüsch*   *Gebüschgesellschaft*
5 Silberweidenwald
6 Grauerlen-Auenwald   } Waldgesellschaften
7 Ulmen-Eschenwald

Das Nebeneinander der Pflanzengesellschaften am Ufer eines Flusses entspricht dem räumlichen Nebeneinander der Standorte. Durch Erosion oder Sedimentation können die Standorte überlagert werden.

Auch hier stehen gesetzmässig Krautflur, Gebüsch und Wald nebeneinander.

(vgl. M. Moor 1958, 1962, 1968 und 1969)

74

Das Gefälle 'Baum-Strauch-Kraut' ist von zwingender Gesetzmässigkeit, gleichgültig ob es sich um extreme Sonderstandorte innerhalb des Waldlandes oder um Rodungen handelt. Sogar in den sehr rasch verlaufenden Sukzessionsreihen auf einer Kahlschlagfläche oder auf durch Windwurf baumfrei gewordenem Waldgelände verläuft die Entwicklung von vorbereitenden Krautfluren über Weichholz-Strauchgesellschaften wieder zum Wald.

Die Entwicklung geht immer von kurzlebiger, z.T. gar ephemerer Krautvegetation über festergefügtes Gebüsch zu langlebigem Wald, von einschichtigen Gesellschaften zu vielschichtigen, von erdnaher Pflanzendecke zu 30 bis 50 und noch mehr Meter hoch aufragender baumhafter Vegetation, von niedrig organisierten, einfach gebauten Pflanzengesellschaften zu den hoch organisierten Wäldern. Die Klimax bilden Pflanzengesellschaften mit den differenziertesten blüten- und verbreitungsbiologischen Verbindungen und Abhängigkeiten, im gemässigten Mitteleuropa die Eichen-Hagebuchenwälder, im westmediterranen Gebiet der Steineichenwald, in äquatorialen Breiten aber der tropische Regenwald mit seinem fast überbordenden Artenreichtum, den nur ihm zukommenden Lebensformen der Epiphyten, Krautstämme und Lianen und einer Fülle von Abhängigkeitsverhältnissen, die heute wohl eher erst erahnt werden als genau bekannt sind.

Regressive Reihen verlaufen vom Wald über Buschwerk zu Krautrasen und in extremen Fällen gar zu vollständig vegetationslosem Gelände. So führten sehr kurze Umtriebszeit, wiederholtes Abbrennen, nachfolgende Beweidung und Übernutzung im westmediterranen Bereich vom Steineichenwald über die Macchia zum Trockenrasen der Garigue, und im Extrem zum vegetationslosen Karstgelände. Die Gebüschformation des Schibljak (*Cotinetalia coggygriae*) stellt ein Degradationsstadium südosteuropäischer Wälder dar. Und im südlichen Mitteleuropa verläuft eine Reihe vom Flaumeichenwald über die Gebüschformation des Berberidion zum Trockenrasen des Xerobrometum.

Man kann das Gefälle 'Baum-Strauch-Kraut' auch grossräumig sehen, so z.B. wenn der Blick vom tropischen Regenwald über Savanne zu Buschland und offener Grassteppe der Passatzonen und über Halbwüste zur Vollwüste der Rossbreiten schweift. Oder vom Galeriewald über den Sträuchermantel zum Grasland des Steppengürtels. Oder vom subalpinen Nadelwald über den Alpenrosen- und Zwergstrauchgürtel zu den Urwiesen der alpinen Stufe und von dort über den lückigen Flechtengürtel der subnivalen Höhenregion zur vegetationsfreien Kältewüste der nivalen Region. Und analog: vom Nadelwald der kühlgemässigten Zone über den Zwergbirkengürtel und die Tundra der subpolaren Gegenden zur Eiswüste der Polkappe.

In allen diesen Fällen ist der Wald durch eine Gebüschzone gegen baumfreies Gelände abgegrenzt. Auf diese Weise entsteht ein allmählicher, weicher, gleitender Übergang, der — in umgekehrter Richtung betrachtet — von der Krautflur über Gebüschgesellschaften zum geschlossenen Wald führt.

Die machtvollste Lebensform im Pflanzenreich ist der Baum. Er trägt seine Erneuerungsknospen hoch über dem Boden, und sie geniessen nur den Schutz, den der Baum mit den Knospenschuppen selbst gewährt. Der Strauch lehnt sich in

Lebens- und Sprossform, Grösse und Lebensdauer an den Baum an, wird aber vom Baum beherrscht und beschattet. Mit seinem Lichthunger findet der Strauch am Rand des schützenden Waldes günstige Lebensbedingungen, tritt dort zu festgefügten Gesellschaften zusammen und bildet das formschöne Bindeglied zwischen Wald und waldfreiem Gelände.

Zusammenfassung

An der Grenze des Waldes gegen baumlose Vegetation haben sich Straucharten zu Gebüschgesellschaften zusammengeschlossen, und zwar sowohl an gehölzfreien Sonderstandorten als auch, halbnatürlich, an den Waldrändern im Rodungsgebiet, so dass der Wald überall durch eine Gebüschzone gegen die Krautflur abgegrenzt erscheint.

Es werden 14 verschiedene Fälle von baumlosen Standorten aufgezählt, natürliche und vom Menschen geschaffene. Überall ist ein gleitender Übergang vom Wald über Gebüsch zur Krautflur festzustellen.

Regressive Reihen, menschlich bedingt, verlaufen ebenfalls vom Wald über Gebüsch zu Rasen, und bei Übernutzung sogar bis zur Vegetationslosigkeit.

Das Gefälle 'Baum-Strauch-Kraut' wird auch grossflächig betrachtet, so vom tropischen Regenwald über Savanne und Buschland zur Steppe oder vom subalpinen Nadelwald über den Zwergstrauchgürtel zu den alpinen Urwiesen.

Literatur

Delelis-Dusollier, A. 1973. Contribution à l'étude des haies, des fourrés préforestiers, des manteaux sylvatiques de France. Diss. Lille.

Dierschke, H. 1974. Die Saumgesellschaften im Vegetations- und Standortsgefälle an Waldrändern. *Script. Geobot.* 6. Göttingen.

Doing, H. 1963. Uebersicht der floristischen Zusammensetzung, der Struktur und der dynamischen Beziehungen niederländischer Wald- und Gebüschgesellschaften. Med. Landb. Hogesch. Wageningen 63.

Ellenberg, H. 1963. Vegetation Mitteleuropas mit den Alpen. Verlag Ulmer, Stuttgart.

Ellenberg, H. 1974. Zeigerwerte der Gefässpflanzen Mitteleuropas. *Script. Geobot.* 9. Göttingen.

Géhu, J.M., J.L. Richard & R. Tüxen 1972. Comte-rendu de l'excursion de l'Association Internationale de Phytosociologie dans le Jura en Juin 1967. Doc. phytosoc. Lille 3.

Hartke, W. 1951. Die Heckenlandschaft. *Erdkunde* 5, 2. Bonn.

Heinrich, W. & R. Marstaller 1973. Uebersicht über die Pflanzengesellschaften der Umgebung von Jena in Thüringen. Wiss. Ztschr. Friedr. Schiller-Univ. Jena, *Math.-Nat. R. Jg.* 22, Heft 3/4.

Hügin, G. 1968. Die Rheinaue im Landschaftsschutzgebiet Taubergiessen. In: Naturschutz und Bildung. Stuttgart.

Jakucs, P. 1959. Ueber die ostbalkanischen Flieder-Buschwälder. *Act. Bot. Acad. Scient. Hung.* 5, 3-4. Budapest.

Jakucs, P. 1970. Bemerkungen zur Saum-Mantel-Frage. *Vegetatio* 21, 1-3.

Jakucs, P. 1972. Dynamische Verbindung der Wälder und Rasen. Akad. Verlag Budapest.

Jurko, A. 1964. Feldheckengesellschaften und Uferweidengebüsche des Westkarpatengebietes. Biologicke prace 10/6. Bratislawa.

Koch, W. 1926. Die Vegetationseinheiten der Linthebene. *Jb. St. Gall. Naturw. Ges.* 61, 2.

Korneck, D. 1974. Xerothermvegetation in Rheinland-Pfalz und Nachbargebieten. Schriftenr. Veget.kde 7. Bonn-Godesberg.

Michalko, J. 1970. Ueber Mantel- und Saumgesellschaften des Verbandes Quercion pubescenti-petraeae. Ber. Int. Symp. Veget. Kde 1966 (Gesellschaftsmorphologie). Den Haag.

Moor, M. 1958, Pflanzengesellschaften schweizerischer Flussauen. *Mitt. schweiz. Anst. forstl. Versuchsw. 34,* 4.

Moor, M. 1960. Waldgesellschaften und ihre zugehörigen Mantelgebüsche am Mückenberg südlich von Aesch (Basel). Bauhinia 1. Basel.

Moor, M. 1969. Zonation und Sukzession am Ufer stehender und fliessender Gewässer. *Vegetatio* 17. Den Haag.

Müller (-Schneider), P. 1977. Verbreitungsbiologie (Diasporologie) der Blütenpflanzen. Veröff. Geobot. Inst. E.T.H., Stiftg Rübel 61, 2. Aufl.

Müller, Th. 1962. Die Saumgesellschaften der Klasse Trifolio-Geranietea sanguinei. *Mitt. Flor.-soz. Arbeitsgem. N.F.* 9.

Müller, Th. 1966. Die Wald-, Gebüsch-, Saum-, Trocken- und Halbtrockenrasengesellschaften des Spitzbergs. In: Der Spitzberg bei Tübingen. Die Natur- u. Landschaftsschutzgebiete Bad.-Württbg. 3. Ludwigsburg.

Müller, Th. 1974. Gebüschgesellschaften im Taubergiessengebiet. In: Das Taubergiessengebiet. Die Natur- und Landschaftsschutzgebiete Bad.-Württbg. 7.

Oberdorfer, E. 1957. Süddeutsche Pflanzengesellschaften. *Pflanzensoziologie* 10. Jena.

Oberdorfer, E. 1970. Pflanzensoziologische Exkursionsflora für Süddeutschland und angrenzende Gebiete. 3. Aufl., Stuttgart.

Oberdorfer, E. 1972. Die synsystematische Gliederung xerothermer Saum-, Busch- und Waldgesellschaften. Beitr. naturk. Forsch. Südw. Dtl. 31. Karlsruhe.

Rameau, J.Cl. & J.M. Royer 1975. Les associations végétales de la Bourgogne calcaire: Diversité et Déterminisme. Comptes rendus du 99e Congrès National des Sociétés savantes (Besançon 1974), Fasc. II. Paris.

Rauschert, S. 1968. Die xerothermen Gebüschgesellschaften Mitteldeutschlands. Diss. Halle (Saale).

Richard, J.L. 1972. La végétation des crêtes rocheuses du Jura. *Ber. Schweiz. Bot. Ges.* 82, 1.

Schwickerath, M. 1944. Das Hohe Venn und seine Randgebiete. Vegetation, Boden und Landschaft. *Pflanzensoziologie* 6. Jena.

Soó, R. 1971. Aufzählung der Assoziationen der ungarischen Vegetation nach den neueren zönosystematisch-nomenklatorischen Ergebnissen. *Act. Bot. Acad. Sc. Hungar.* 17. Budapest.

Troll, C. 1951. Hecken im maritimen Grünlandgürtel Mitteleuropas. *Erdkunde* 5. Bonn.

Tüxen, R. 1952. Hecken und Gebüsche. *Mitt. Geogr. Ges. Hamburg* 50.

Tüxen, R. 1955. Das System der nordwestdeutschen Pflanzengesellschaften. *Mitt. Flor.-soz. Arbeitsgem. N.F.* 5.

Tüxen, R. & E. Oberdorfer 1958. Die Pflanzenwelt Spaniens. Veröff. Geobot. Inst. Rübel Zürich 32, Teil II (Eurosib. Phan. Ges. Spaniens).

Wendelberger, G. 1967. Grundzüge zu einer Vegetationskunde Salzburgs. *Mitt. Oesterr. Geogr. Ges.* 109, 1-3.

Westhoff, V. & A.J. Den Held 1969, Plantengemeenschappen in Nederland, Zutphen.

Adress of the Author:
Dr. M. Moor
Hohe Winde-Str. 19
CH - 4059 Basel, Schweiz

SIGMETEN UND GEOSIGMETEN, IHRE ORDNUNG UND IHRE BEDEUTUNG FÜR WISSENSCHAFT, NATURSCHUTZ UND PLANUNG

Reinhold Tüxen

Einleitung

Alexander von Humboldt hat gelehrt, daß die Pflanzendecke, das Pflanzenkleid, die Eigenart einer Erdgegend bestimme. 'Kleider machen Leute'. Pflanzen-Kleider sind aber weit mehr als äußerliche Merkmale der Landschaft; sie sind − mit Einschluß der meist weniger augenfälligen Tiere − der umfassende Ausdruck aller physischen Lebensmöglichkeiten und ihrer Geschichte.

Weder Pflanze noch Tier noch Mensch leben für sich allein, sondern vom Beginn ihres Daseins bis zu ihrem Ende in Gesellschaften. Raymond Rapaics (1931), der dieses allgemeine Grundgesetz erneut in aller Klarheit aussprach, hob zugleich hervor, 'daß auch das Verhältnis der Gesellschaft zur Gesellschaft nicht als ungeregelt betrachtet werden kann, und daß auch dieses Verhältnis seine Naturgesetze hat.'

In seiner 'Allgemeinen Vegetationsgeographie' prägte Josef Schmithüsen (1959, p. 164, 1961, p. 164, 1968, p. 303) den Begriff des 'landschaftlichen Vegetationskomplexes', d.h. des 'für eine Gegend charakteristischen Gesellschafts-i n v e n t a r s.' Er spornte damit Pflanzensoziologen, die sich zunächst mit dem Studium der A r t e n -Verbindungen und ihrer Auswertung in verschiedenen Richtungen beschäftigt hatten, zur planmäßigen Aufnahme der Gesellschaftsausstattung in natürlichen Vegetationsgebieten steigender Größenordnung an.

Schon früher war von Assoziationskomplexen gesprochen worden (vgl. Gams 1918, p. 446). Braun-Blanquet & Pavillard (1928) faßten unter dem Einfluß der Mono-Klimax-Theorie 'die Gesamtheit aller Entwicklungsserien, die einer bestimmten klimatischen Schlußgesellschaft zusteuern' als Klimax-Komplex zusammen, der auch als Synoekosystem bezeichnet wurde (Braun-Blanquet 1964, p. 664). Friedel (1956) behandelte die Gesellschaftskomplexe im Hochgebirge und ihre Kartierung.

Der Vergleich der zunächst noch sehr unvollständig bekannten Gesellschafts-Verbindungen in natürlichen Vegetationsräumen nach dem Muster der Gegenüberstellung von Artenlisten der Pflanzengesellschaften ergab schon frühzeitig Trenn- und Kenngesellschaften für die einzelnen Gebiete. Sie wurden zur Benennung der Vegetations-Landschaften verwendet und bürgerten sich rasch ein: Eichen-Birkenwald-, Eichen-Hainbuchenwald-, Erlenbruchwald- usw. Landschaften oder -Gebiete (R. Tüxen 1931, 1934, 1939 a,b, 1942, 1951, 1956, 1957, Schmithüsen 1940, Knapp 1948, 1958, 1971, Krause 1950, 1952, 1955, Braun-Blanquet & R. Tüxen 1952, R. Tüxen & Oberdorfer 1958).

Ein Vorschlag zur planmäßigen Aufnahme solcher Gesellschafts-Inventare und zur Fassung ihrer Typen wurde als Auswirkung der Anregung von Josef Schmithüsen aber erst 1973 (R. Tüxen) vorgelegt, der sofort in der Schweiz (Béguin), in Frankreich (J.M. Géhu) und in Spanien (S. Rivas Martinez) aufgegriffen wurde.

Die Methode der Aufnahme von Gesellschafts-Komplexen wurde inzwischen von Wilmanns und Tüxen (1978) verfeinert.

Ähnliche Gedanken entwickelte Rüdiger Knapp 1975. Er prägte den Ausdruck Geo-Syntaxon für Gesellschaftskomplexe als Raumeinheiten nach dem folgenden Schema, worin er die formale Parallelität aufzeigte:

Als Objekt der Klassifizierung in einem System		Entsprechender Wissenschaftszweig
Sippe (Art, Subspecies)	------- Taxon	Taxonomie
Pflanzen- bzw. Tier-Gesellschaft	------- Syntaxon	Syntaxonomie
Gesellschaftskomplexe als Raumeinheiten	------- Geosyntaxon	Geosyntaxonomie[1]

Nach weiteren meist aufeinander abgestimmten Untersuchungen in verschiedenen Gebieten konnte 1977 der Problemkreis der Assoziationskomplexe zum Thema des 21. Symposion der Internationalen Vereinigung für Vegetationskunde gemacht werden, das die Ergebnisse zahlreicher Bearbeiter aus Japan, Europa (Polen, Deutschland, Schweiz, Italien, Frankreich, Spanien) und eine vorläufige Übersicht über die Küsten-Dünen-Vegetation der Holarktis von Canada bis Japan vorlegte (cf. R. Tüxen 1978). Hier begannen sich an weit gestreuten Beispielen, aus verschiedenen Blickrichtungen betrachtet und von Fernerstehenden und Skeptikern kritisch beleuchtet, Begriffe, Terminologie, Arbeitsmethoden und Anwendungsbereiche zu klären, sodaß die weitere Entwicklung der Synsoziologie kraftvoll voranschreiten wird.

Um dazu, zugleich als Dank für die entscheidende Anregung des Jubilars, weiter beizutragen, seien hier einige inzwischen gewonnene Einsichten in die Möglichkeiten zur Systematisierung der Geosyntaxa, sowie zu ihrer Anwendung in verschiedenen Bereichen mitgeteilt.

1. Sigmetum

Als Sigmetum (Synonym: Sigmassoziation) fassen wir die unterste Einheit von Geosyntaxa, d.h. im Gelände regelmäßig wiederkehrende Vergesellschaftungen von Pflanzengesellschaften (Phytozönosen, Biozönosen) auf, die noch durch eine oder mehrere Kenn-Gesellschaften ausgezeichnet sind. Ihr Typus wird (wie derje-

---

1 Von Rivas Martìnez (1976) als Synsoziologie bezeichnet.

nige der Assoziationen) in einer Tabelle dargestellt, die statt der Arten (Taxa) die Gesellschaften (Syntaxa) enthält. An ihr werden die Reinheit und die Möglichkeit zur Gliederung in Untereinheiten (Subsigmeten usf.) mit Hilfe von Trenngesellschaften geprüft. Die hierzu verwendete Arbeitsweise ist dieselbe wie bei der Anfertigung der Assoziationstabellen.

Während des 21. Internationalen Symposion 1977 in Rinteln wurde der Terminus 'Sigmetum' als handliche Abkürzung für die vorher gebrauchte Bezeichnung Sigmassoziation vorgeschlagen und bürgerte sich rasch ein.

Ebenso, wie die Assoziation mit Hilfe der Tabelle objektiv durch ihre Artenkombination charakterisiert wird, d.h. also ein Abstraktum darstellt, wird auch das Sigmetum durch die Tabelle ihrer Gesellschaftskombination und den daraus sich ergebenden synthetischen Merkmalen (Treue, Stetigkeit, Gesamtzahl der Gesellschaften, notwendige Aufnahmezahl) sowie durch die mittlere Aufnahmezahl (analytisches Merkmal) gekennzeichnet.

Die Benennung der Sigmeten erfolgt – wie diejenige der Assoziationen nach dominierenden und Kenn-Arten – nach vorherrschenden Kenn-Gesellschaften (z.B. S a l i c e t o   a l b a e - S i g m e t u m).

Die Gesetze, die für das Zusammenleben der Taxa (Pflanzen und Tiere) in Gesellschaften gelten (R. Tüxen 1965) sind sinngemäß auch für die Vergesellschaftung der Syntaxa (Pflanzengesellschaften, Biozönosen) zu Sigmeten anwendbar und auszuwerten.

1. Gesetz der Gesellschaftsordnung der Lebewesen,
2. Gesetz der exogenen Ordnung der Lebensgemeinschaften (Biozönosen),
3. Gesetz der räumlichen Ordnung,
4. Gesetz der zeitlichen Ordnung,
5. Gesetz der endogenen (funktionalen) Ordnung,
6. Gesetz der Produktionsordnung,
7. Gesetz der Harmonie.

Syntaxonomisch verwandte Sigmeten werden durch verbindende Kenn-Gesellschaften (nicht nur Assoziationen, sondern auch deren Fragmente oder andere ranglose Syntaxa) zu Verbänden (Sigmion), diese zu Ordnungen (Sigmetalia) und endlich diese zu Klassen (Sigmetea) vereinigt (Abb.). Dieses abstrakte System der Sigmeten als dem niedrigsten Komplexitätsgrad der Geosyntaxa entspricht also dem der Assoziationen (Syntaxa). Seine höheren Einheiten enthalten – ebenso wie jene der Syntaxa alle Arten – alle zugehörigen Gesellschaften (Biozönosen) und damit auch alle in ihnen lebenden Arten. Es stellt also eine höhere Integrationsstufe mit einem umfassenderen Inhalt und zugleich mit weiter und tiefer reichenden Deutungs-, Auswertungs- und Anwendungsmöglichkeiten dar, zumal alle Eigenschaften, Ursachen und Wirkungen der darin vereinigten Gesellschaftsgruppen, wie deren Struktur (Symmorphologie), Syndynamik, Synchorologie, Synchronologie, Synökologie und Symproduktion darin enthalten sind.

Das zeigen auf den ersten Blick schon die bisher veröffentlichten Beispiele solcher höherer Einheiten des geosyntaxonomischen Systems, z.B. von J. Tüxen,

J.M. Géhu, A. Miyawaki, R. Tüxen, K.H. Hülbusch und D. Kienast (s. R. Tüxen 1978).

Nach ihrer Entstehung und Abhängigkeit von exogenen Faktoren unterscheiden wir drei Gruppen von Sigmeten, die durch alle denkbaren Übergänge miteinander verbunden sein können. (R. Knapp (1975, p. 403) kam nach der unterschiedlichen Dynamik und der anthropogenen Beeinflussung zu vier anderen Hauptgruppen von Gesellschaftskomplexen, auf die wir verweisen möchten.)

1. Primär-Sigmeten sind Vergesellschaftungen von natürlichen oder doch sehr naturnahen Syntaxa (Biozönosen), wie sie im unberührten Hochgebirge, in der Arktis und Antarktis, in ungestörten tropischen Regenwaldgebieten, an entlegenen Meeresküsten, Seen, Flüssen oder in lebenden Hochmooren noch bestehen dürften. Die Anwendung des Begriffes der 'potentiellen' natürlichen Vegetation erübrigt sich hier, weil die natürlichen Vegetationseinheiten real vorhanden sind.

2. Sekundär-Sigmeten werden aus Ersatzgesellschaften 1. und 2. Grades gebildet, denen auch natürliche und naturnahe Syntaxa beigesellt sein können. Die anthropo-ökologisch (auch historisch) bedingte strukturelle Spanne dieser Gruppe reicht sehr weit. Sie umfaßt alle Wirtschafts-Landschaften, in denen der Boden in seinen Oberflächenformen nicht plötzlich irreversibel vom Menschen verändert worden ist. Diese Gebiete enthalten neben Resten natürlicher Gesellschaften durch anthropo-zoogene Einflüsse (Brand, Weide, Mahd, mäßige Düngung, oberflächliche Ent- und Bewässerung, Acker-Bestellung, Anbau von Feldfrüchten und gesellschafts- und standortfremden Holzarten) erzeugte Ersatzgesellschaften. Die durch diese 'Frisur' der natürlichen Vegetation (und auch durch stärkere Eingriffe) entstehende Umformung derselben bringt zwangsläufig Veränderungen des Lokalklimas, des Wasserhaushaltes, der Struktur des Bodens und damit seiner biologischen Aktivität mit sich, die zunächst reversibel sind, jedoch zu irreversiblen Folgen führen können. Daher rechnen wir auch die Gebiete, in denen alte Erosion, Auelehmbildung, Podsolierung und ähnliche vom Menschen ausgelöste und langsam fortschreitende Vorgänge die ursprüngliche Vegetation geändert haben, noch zu den sekundären Sigmeten.

Diese Beispiele erklären die weite Spanne der Sekundär-Sigmeten und zugleich die Schwierigkeit ihrer Abgrenzung gegen die Primär- und Tertiär-Sigmeten. Sekundär-Sigmeten liegen im Bereich derselben oder nahe verwandten Einheiten der potentiellen natürlichen Vegetation.

3. Tertiär-Sigmeten (Ersatz- oder Substitutions-Sigmeten) sind Gesellschaftskomplexe auf künstlich vom Menschen stark veränderten oder neu geschaffenen Standorten wie Siedlungen aller Art: Einzelhäuser bis Großstadt, Industrie- und Verkehrsanlagen u.a. Hier wird der Begriff der potentiellen natürlichen Vegetation wiederum um so weniger anwendbar, je krasser die Eingriffe und damit die durch sie bewirkten Veränderungen des ursprünglichen natürlichen Komplexes abiotischer Standortseigenschaften sind. Die potentielle natürliche Vegetation könnte zwar für einen Zeitraum, der vor demjenigen liegt, welcher die heutigen Lebensbedingungen entscheidend verändert hat, erschlossen werden. Dies hat aber mit dem

82

derzeit vorhandenen Lebens-Potential nichts mehr zu tun. Im äußersten Falle ist vielmehr nur noch die an die künstlich vom Menschen geschaffenen Standorte angepaßte Vegetation möglich. Eine andere, nach dem Aufhören des menschlichen Einflusses sich spontan einstellende Vegetation ist aber für die Beurteilung der heute herrschenden Lebensmöglichkeiten gleichgültig.

Nach der Stärke der menschlich bedingten Standorts-Veränderung lassen sich Tertiär-Sigmeten verschiedenen Grades unterscheiden:

a. Alte, weiträumige Siedlungen mit bodenständigen Gehölz-Pflanzungen, z.B. nw-deutsche Streusiedlungen mit Eichen-Kämpen.

b. Geschlossene Haufendörfer, Teile alter kleinstädtischer Siedlungen mit Gemüse- und Obstgärten.

c. Neue Villenviertel mit Zierrasen, Blumengärten und ausländischen Gehölzen (Koniferen) u.a.

d. Standorte mit künstlichem, tiefgreifendem Ab- und Auftrag von Boden und Gestein (abgetorfte Hochmoore, Bergwerkshalden, Spülflächen).

e. Fabrikanlagen und Großstadt-Kerne mit vollständiger Bedeckung der Erd-oberfläche durch lebensfeindliche Substrate (Ziegel, Asphalt, Beton usw.)

Tertiär-Sigmeten sind die Anfänge der Ansiedlung und Entwicklung von Biozönosen auf künstlich geschaffenen Substraten und ähneln damit gewissermaßen 'Prothesen' in der natürlichen Landschaft. Sie haben keine Beziehung zum natürlichen Ausgangszustand. Ihre Entwicklung nach dem Aufhören der sie schaffenden und erhaltenden Einflüsse führt zu 'Ausheilungsstufen' in Richtung auf eine naturnähere Vegetation, welche aber nie die ehemalige, ursprüngliche Ausbildung erreichen können. Beispiele von Tertiär-Sigmeten wurden von Hülbusch (1977) und Kienast (1977) an Hand der Vegetation der Stadt Schleswig dargestellt (Vgl. R. Tüxen 1978).

Alle Sigmeten, von denen zahlreiche Beispiele in dem Symposion-Bericht mitgeteilt wurden, und auf die wir hier verweisen müssen, sind am sichersten an ihrer Gesellschafts-Kombination und innerhalb derselben vor allem an ihren Kenngesellschaften, für den mit ihnen Vertrauten aber auch an ihrer Physiognomie leicht zu erkennen, so daß sie aus dem fahrenden Kraftwagen oder Zuge ebenso angesprochen werden können, wie aus der Luft (Luftbilder).

2. Geosigmetum

Wie auf engem Raum in den Gesellschafts-Beständen die Individuen der Taxa und auf größerer Fläche in einem Sigmetum die Bestände der Syntaxa nur beschränkte, bestimmte Kontakt-Möglichkeiten haben, so grenzen einzelne Sigmetum-Flächen immer nur an solche bestimmter anderer Sigmeten. So wenig, wie *Lemna gibba* mit *Bromus erectus* oder *Arum maculatum* in engem Kontakt leben leben kann, so wenig, wie ein Bestand des N u p h a r e t u m an einen des C o r y - n e p h o r e t u m c a n e s c e n t i s oder eines L e u c o b r y o - P i n e t u m stoßen kann, ebenso wenig berührt eine Fläche eines A m m o p h i l e t o - S i g -

m e t u m eine solche des A l n e t o i n c a n a e - S i g m e t u m oder des P i n e t o m u g i - S i g m e t u m.

Folglich können auch bestimmte regelmäßig wiederkehrende gesetzmäßige räumliche Vergesellschaftungen von Sigmeten unterschieden, definiert und kartiert werden (Abb.).

Die nächst höhere Stufe der räumlichen Vergesellschaftung von Sigmeten wurde Geo-Sigmetum genannt (Tüxen, R. 1978, p. 11). Ihr Typus ist durch ein oder mehrere Kenn-Sigmeten ausgezeichnet. (Zoller, Béguin & Hegg 1978 brauchen die Begriffe Geosyntaxon und Geosigmetum, die auf Gesellschaftsverbänden beruhen und das daraus aufgebaute syntaxonomische System (Geosigmion, Geosigmetalia), wie es scheint, in einem abweichenden Sinn.)

Das Wort Geosigmetum nimmt vielleicht zu wenig Rücksicht auf den umfassenden Begriff des Geosyntaxons (Knapp 1975), bezeichnet es doch in dem hier gebrauchten Sinne nur einen bestimmten Komplexitätsgrad. Es fragt sich daher, ob es beibehalten werden kann oder durch einen anderen Terminus ersetzt werden muß. Es könnte jedoch als untere Stufe der Sigmetum-Gesellschaften von geschlossener Flächenbedeckung Berechtigung finden.

Im Flachland sind die Areale der Geosigmeten sehr ausgedehnt, während sie im Gebirge oder an den Küsten viel enger werden und mit dem klimatisch oder geomorphologisch bedingten Standortgefälle (Gradient) dichter gebündelt werden können. Schmithüsen wies in seiner 'Allgemeinen Vegetationsgeographie' bereits auf solche 'Standortsreihen' (Gruppen, Catenen) hin (1959, p. 126 f., 1968, p. 240 f.).

Diese Kontaktgruppen sind aber zunächst ebensowenig Typen von Einheiten unseres geosyntaxonomischen Systems wie die natürlichen Kontakte zweier Assoziationen, z.B. F a g e t u m und S e s l e r i e t u m, C a r p i n o - Q u e r c e t u m und C a r i c i e l o n g a t a e A l n e t u m eine syntaxonomische Verwandtschaft bedeuten.

Als Beispiele von Geosigmeten seien die räumliche Vergesellschaftung der folgenden Sigmeten genannt:

Im Altmoränen-Gebiet NW-Deutschlands (B e t u l o - Q u e r c e t o - G e o - s i g m e t u m finden sich regelmäßig vereinigt: B e t u l o - Q u e r c e t o - S i g - m e t u m, S p h a g n e t o m a g e l l a n i c i - S i g m e t u m, C a r i c i e l o n - g a t a e - A l n e t o - S i g m e t u m, V i o l o - Q u e r c e t o - (F a g o - Q u e r - c e t o -) S i g m e t u m, (P o t a m e t a l i a) - S i g m e t u m Seen), S c i r p o - P h r a g m i t e t o - S i g m e t u m, P o l y g o n e t o b r i t t i n g e r i - S i g - m e t u m, S a l i c e t o t r i a n d r o - v i m i n a l i s - S i g m e t u m, F r a x i - n o - U l m e t o - S i g m e t u m, Q u e r c o - C a r p i n e t o - S i g m e t u m, diverse Siedlungs-Sigmeten u.a.

Im Niedersächsischen Hügelland (C a r p i n o - Q u e r c e t o (= Q u e r c o - C a r p i n e t o -) G e o s i g m e t u m) sind vereinigt: Q u e r c o - C a r p i n e - t o - S i g m e t u m, F r a x i n o - U l m e t u m - S i g m e t u m, S a l i c e t o t r i a n d r o - v i m i n a l i s - S i g m e t u m, C a r i c i e l o n g a t a e - A l n e t o - S i g m e t u m, R i b o s y l v e s t r i s - A l n e t o - S i g m e t u m,

Stellario-Alneto-Sigmetum, (Potametalia)-Sigmetum, Scirpo-Phragmitetum-Sigmetum, Polygoneto brittingeri-Sigmetum, diverse Siedlungs-Sigmeten u.a.

Die eingeklammerten Namen sind provisorisch. Sie sollen nur die Anwesenheit eines noch nicht genauer untersuchten und benannten Sigmetum anzeigen.

Geosigmeten springen bei guter Farbgebung und ausreichend feiner Legende aus den Karten sowohl der realen als auch vor allem der potentiellen natürlichen Vegetation nicht zu großen Maßstabes geradezu ins Auge.

So zeigt die neue Karte der 'Potentiell natürlichen Pflanzendecke Niedersachsens' (Der Niedersächs. Minister f. Ernährung, Landwirtschaft u. Forsten) sehr deutlich, (ohne damit zu der Begrenzung der einzelnen Einheiten Stellung nehmen zu wollen), folgende Geosigmeten, die zum großen Teil übrigens schon in den Vorläufern (Tüxen 1934, 1939) dieser viel eingehenderen Karte angedeutet sind:

1. Ammophileto arenariae-Geosigmetum, 2. Puccinellio maritimae-Geosigmetum (Flußaufwärts wohl zu weit gefaßt), 3. Betulo-Querceto-Geosigmetum, 4. Carpino-Querceto-(=Querco-Carpineto-)Geosigmetum, 5. Fageto-Geosigmetum, 6. Piceato-Geosigmetum.

(Die Benennung dieser Einheiten ist noch provisorisch und kann erst dann endgültig geprägt werden, wenn Tabellen der Sigmetum-Kombinationen von den einzelnen Geosigmeten vorliegen).

Eindrucksvolle bildliche Darstellungen von Sigmeten, aber auch von Geosigmeten sind in geographischen Bildwerken (z.B. *Hamaya* 1975) zu finden, wenn sie auch nicht als solche ausgewertet wurden.

Von den große bis sehr große Oberflächen bedeckenden Sigmetum-Gesellschaften, den Geo-Sigmeten, lassen sich – vorausgesetzt, daß man die Sigmeten kennt – sei es im Gelände mit Hilfe von Transekten oder an Hand von genügend genauen Vegetationskarten 'Aufnahmen', d.h. Listen der auf einem ausreichend großen einheitlichen Minimal-Raum vorkommenden Sigmeten mit Schätzung ihres Flächenanteils gewinnen und daraus Tabellen anfertigen. Diese ergeben die Sigmeten-Kombinationen und durch deren Vergleich die Kenn-Sigmeten der einzelnen Typen. Deren Untereinheiten (Sub-Geosigmeten) können durch Trenn-Sigmeten gekennzeichnet werden.

Geosigmeten sollten nach ihren verbreitetsten Kenn-Sigmeten benannt werden. (Verbände oder Ordnungen der Syntaxonomie, wie Fagion, Quercetalia pubescentis usw. dürften dafür nicht geeignet sein.)

Ebenso wie sich syntaxonomisch verwandte Sigmeten auf Grund von verbindenden Kenn-Syntaxa zu Verbänden und höheren Einheiten zusammenfassen lassen, dürften auch syntaxonomisch verwandte Geosigmeten mit Hilfe von Kenn-Sigmeten zu Verbänden zu gruppieren sein (Geosigmion). So zeigen gewiß die durch Quercion robori-petraeae-Gesellschaften gekennzeichneten Geo-Sigmenten von Galizien (NW-Spanien), der Landes (SW-Frankreich), der westlichen Bretagne, SW-Irlands, der Niederlande und NW-Deutschlands manche Gemeinsamkeiten.

Weil aber mit der räumlichen Ausdehnung der Geosigmeten, d.h. also der Vergesellschaftung der Sigmeten, ihre Eigenständigkeit (Singularität) durch den Wechsel der klimatischen, geologischen, pedologischen, hydrologischen und historischen Einflüsse zunimmt, wird es wahrscheinlich, daß sich höhere syntaxonomische Kategorien wie Ordnungen oder gar Klassen durch die Verbindung auf Grund von charakteristischen Sigmeten ergeben werden. Es wäre allerdings daran zu denken, nach dem Muster der Klassen-Gruppen in der Syntaxonomie für die Zusammenfassung verwandter Geosigmeten oder ihrer Verbände (Geosigmion) statt verbindender Sigmeten das gemeinsame Vorkommen von Sigmion-Verbänden zu nutzen.

Man könnte endlich prüfen, ob die Verbände (oder Ordnungen) der Syntaxa (z.B. Q u e r c i o n r o b o r i - p e t r a e a e, S p h a g n e t a l i a p a p i l l o s i, A r n o s e r i d i o n) sich als Binde-Glieder höherer Rangstufen von Geo-Syntaxa eignen. Diese Frage kann aber erst entschieden werden, wenn Tabellen solcher höherer Einheiten vorliegen.

Eindrucksvolle physiognomische Ähnlichkeiten lassen sich auch von analogen Geosigmeten in verschiedenen Erdteilen aufzeigen, die auf Formations-Merkmalen beruhen, die aber wie diese keine Übereinstimmung der Taxa, der Syntaxa oder Geosyntaxa zeigen. Man könnte sie (nach einem Vorschlag von Frau Prof. Otti Wilmanns) als Iso-Sigmeten (Iso-Geosigmeten) bezeichnen. Auch diese würden, besonders in der Geographie, weitere Betrachtung verdienen.

3. Höhere räumliche Komplexitätsgrade oder geosyntaxonomische Einheiten

Die in kleinerem Karten-Maßstab deutlich werdende, auf einer gewissen Fläche als Mosaik wiederkehrende gesetzmäßige Vergesellschaftung von Geosigmeten, wie sie das Zusammenspiel von Allgemein-Klima, Geologie und Relief bei gleicher Vegetations-Geschichte erzeugt, kann zu einer noch höheren Stufe in der Hierarchie der Geosigmeten zusammengefaßt werden. Jeder Typus dieser Stufen wird durch Kenn-Geosyntaxa der nächst tieferen Stufe gekennzeichnet. Als Beispiel verweisen wir auf die gesetzmäßige Anordnung der 'Q u e r c i o n r o b o r i - p e t r a e a e -', 'F r a x i n o - C a r p i n i o n'- und 'F a g i o n'-Geosigmeten[1] in der atlantischen Provinz West-Europas oder ihr Gegenstück in den 'P i n i o n s y l v e s t r i s'-, 'Q u e r c e t a l i a p u b e s c e n t i s'- und 'C a r p i n i o n'-Geosigmeten Osteuropas oder auf die gesetzmäßigen Höhenstufen-Folgen der Hochgebirge. Schließlich lassen sich noch eine (oder mehrere) weitere Ränge solcher räumlicher Kontakt-Mosaiken auch dieser Einheiten erkennen, die sich zuletzt mit den höchsten pflanzengeographischen Räumen decken. Man könnte die oberste Stufe dieser Rangordnung als Holo-Sigmetum bezeichnen (Abb.). (Für die darunter folgenden müßten geeignete Präfixe gefunden werden.)

---

1 Diese Bezeichnungen sind provisorisch und haben keine nomenklatorische Gültigkeit!

4. Geosyntaxonomische Systeme

Die Typen der Geosyntaxa lassen sich also in zweifacher Weise hierarchisch ordnen.

1. Wie im syntaxonomischen System werden soziologisch verwandte Geosyntaxa nach ihrer soziologischen Struktur – 'Konstitution' – durch verbindende Kenn-Syntaxa – ohne Rücksicht auf ihre räumliche Lage und Größe – zu Verbänden, Ordnungen und Klassen vereinigt (s.p. 81 u. Abb.). Dieses System ist eine Abstraktion, wie alle taxonomischen und syntaxonomischen Einheiten und Rangstufen.

2. Der gesetzmäßige Kontakt der Geosyntaxa läßt Typen von räumlich vorhandenen Geosyntaxa-Kombinationen erkennen, die auf Grund gemeinsamer geosyntaxonomischer Einheiten zu höheren Rängen solcher Vegetationsgebiete zusammengefügt werden können (Abb.).

In beiden Systemen, dem 'syntaxonomischen' und dem 'geographischen' wird jede höhere Stufe durch eine selbständige Geo-Syntaxa-Kombination mit eigenen Kenn-Einheiten der nächst tieferen gekennzeichnet. Für die Begrenzung der abstrakten Einheiten der Geosyntaxa-Kombinationen ist also innerhalb dieser das Vorhandensein oder Fehlen von Kenn-Einheiten der nächst tieferen Stufe entscheidend. Die Geosyntaxa-Kombinationen mit ihren Kenn- und Trenn-Syntaxa der verschiedenen Grade des geographischen Geosyntaxa-Systems sind an bestimmte Erdräume von steigender Größe gebunden. Auch sie werden nach dem Auftreten oder Fehlen von Kenn-Einheiten der nächst tieferen Stufe innerhalb der Kombinationen von Geosyntaxa in deren Verbreitungsgebieten gegeneinander abgegrenzt.

Die hierarchischen Stufen beider Systeme decken sich also durchaus nicht. Das geosyntaxonomische System stuft also alle bekannten Geosyntaxa sozusagen in 'vertikaler' Rangordnung nach ihrer soziologischen Verwandtschaft ohne Rücksicht auf ihre geographische Verbreitung ein. Das 'geographische' System ordnet die Typen der Vergesellschaftung von Geosyntaxa gewissermaßen horizontal zu einer Hierarchie von Vegetationstypen nach Erdräumen wachsender Größe.

Ebenso wie die Individuen bestimmter taxonomischer Einheiten, z.B. einer Gattung (z.B. *Ranunculus*), in sehr vielen Assoziationen, Verbänden, Ordnungen und Klassen oder dieselben syntaxonomischen Einheiten (Assoziationen) in zahlreichen geosyntaxonomischen Typen verschiedener Rangstufen vorkommen, finden sich die gleichen geosyntaxonomischen Einheiten (z.B. Sigmeten) in mehreren Geosigmeten oder deren höheren hierarchischen Stufen. Das gilt für beide Systeme. Das folgende Schema möge das Verhältnis der Taxa, Syntaxa und Geosyntaxa und ihrer Systeme deutlich machen (Abb.).

5. Anwendungsbereiche

Die Verwendungsmöglichkeiten der hier dargestellten Begriffe und Einheiten in Wissenschaft und Planung sowie im Naturschutz sind vielseitig und können sehr fruchtbar werden.

Zunahme der taxonomischen und geosyntaxonomischen Vielfalt

Vegetationsgeographische Einheiten

Holo-Sigmetum

"Meso"-Geosigmetum?

Geosigmetum   Geosigmion?

Sigmetum-Sigmion-Sigmetalia-Sigmetea

Assoziation-Verband-Ordnung-Klasse-Klassengruppe

Species-Genus-Familie-Ordnung-Klasse-Abteilung

Taxonomische, syntaxonomische und geosyntaxonomische Einheiten

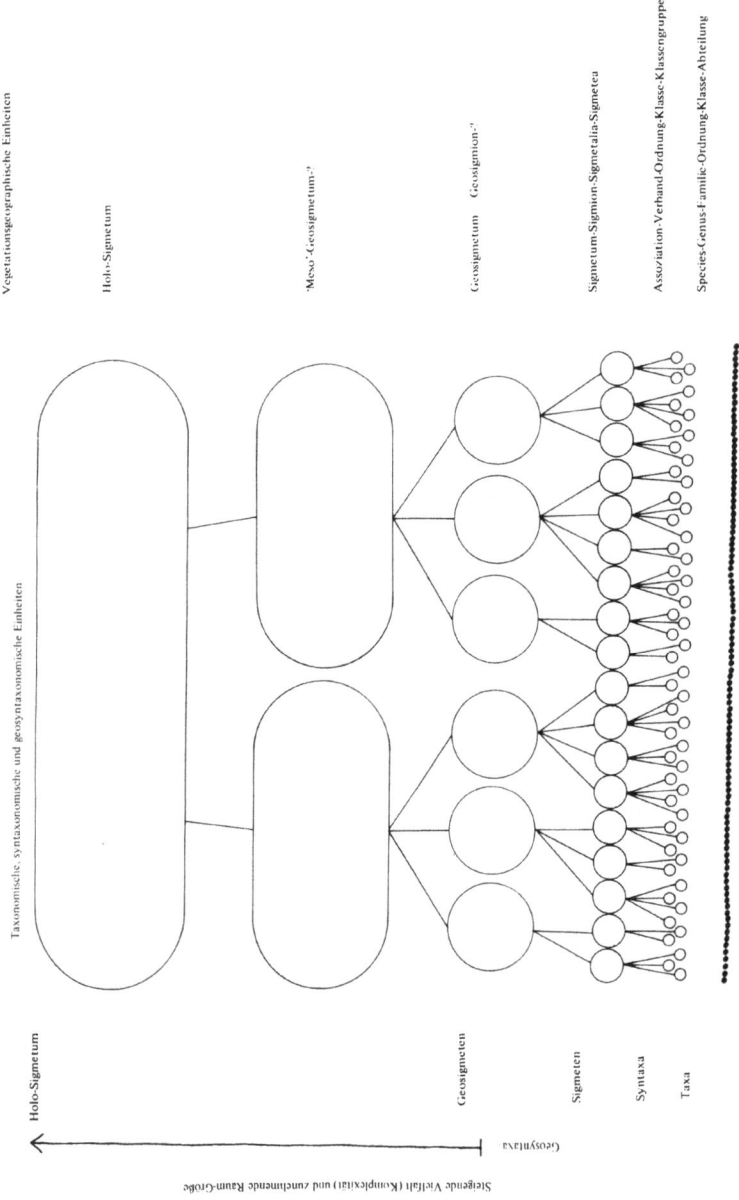

Holo-Sigmetum

Geosigmeten

Sigmeten

Syntaxa

Taxa

(Geosyntaxa)

Steigende Vielfalt (Komplexität) und zunehmende Raum-Größe

1. **Pflanzengeographie.** Die Bedeutung der Geosyntaxa (Sigmeten, Geosigmeten und ihrer höheren räumlichen Integrationsstufen) für die scharfe Begrenzung der pflanzengeographischen Raum-Einheiten wurde schon von Josef Schmithüsen (1959, p. 164, 1968, p. 303) klar vorausgesehen. 'Raumeinheiten der höheren Rangstufen (Bezirke, Provinzen, Kreise, Reiche) sind jeweils aus mehreren aneinandergrenzenden nach ihrem Gesellschaftsinventar verwandten Gebieten der niederen Rangstufen zusammengesetzt. Sie lassen sich durch gebietseigene Gesellschaftseinheiten kennzeichnen.'

Dafür wäre die Bindung der geosyntaxonomischen Einheiten an die pflanzengeographischen Sektionen, Provinzen und Regionen usf. (vgl. z.B. Braun-Blanquet 1964, p. 750) zu untersuchen. Wenn sie besteht – und vieles scheint dafür zu sprechen – dürften gerade die Geosyntaxa und ihre Rangstufen die bisher schwer festzulegenden Grenzen der pflanzengeographischen Räume viel schärfer fassen können.

2. **Vegetationskartierung.** Durch die Verwendung von Sigmeten, Geosigmeten und der höheren geosyntaxonomischen Stufen des räumlichen Systems der Geosyntaxa als Kartierungseinheiten kann unter einem Begriff, also in der Karte mit einer Signatur die gesamte reale Vegetation (und die Flora) großer und sehr großer Gebiete dargestellt werden. Dadurch wird die Kartierung der realen Vegetation ganzer Kontinente, ja der Erde, sicher, scharf und vollständig möglich. Denn diese Einheiten sind die Gesamtheit, das Integral aller Arten (Pflanzen und Tiere) einschließlich aller Kulturpflanzen und angebauten, auch der gebietsfremden Holzarten, sowie der realen und auch der potentiellen natürlichen Vegetation und damit zugleich der vollständige Ausdruck aller wirtschaftlichen Gegebenheiten und Möglichkeiten und des Ertrags-Potentials.

Die Grenzen der geosyntaxonomischen Einheiten werden die naturräumliche Gliederung der Erdoberfläche schärfer und sicherer erkennen lassen.

3. **Tiersoziologie.** Die Sigmeten (und Geosigmeten) liefern eine fruchtbarere Grundlage und ein schärfer auswertbares Bezugssystem für die Tier-Geographie und -Soziologie als die Pflanzengesellschaften (Syntaxa) allein, weil viele Tiere nicht auf eine Pflanzengesellschaft beschränkt, sondern eher mit Sigmeten oder gar mit Geosigmeten verbunden sind.

4. **Pedologie, Geologie, Hydrologie.** Ähnlich aufschlußreich wird der Vergleich der Boden-Gesellschaften (Catenen) mit der Karte der Sigmeten und Geosigmeten, der auf den Bereich der Geologie und Hydrologie ausgeweitet werden könnte.

5. **Landschaftsökologie.** Die Synsoziologie kann als empirische Arbeitsweise der Landschaftsökologie gesicherte, scharf definierte Grundlagen zur Verfügung stellen, die es ermöglichen, Austausch-Vorgänge und gegenseitige Beeinflussung (Epharmonie) der wenig einheitlich und scharf begrenzten 'Ökosysteme' (vgl. Schwabe 1977) klarer zu fassen.

6. **Naturschutz.** Die Erkenntnis größerer räumlicher gesetzmäßiger Zusammenhänge der Pflanzengesellschaften (Biozönosen), Sigmeten und Geosigmeten wird dem Naturschutz neue Einsichten und Ziele geben.

7. **Geographie.** Aus allem ergibt sich die Bedeutung der geo-syntaxonomischen

Einheiten für die Geographie, die auf der gesicherten Kenntnis der Arten-(Taxa), Gesellschaften (Syntaxa), Geosyntaxa (Sigmeten, Geosigmeten usw.) beruht.

Damit dürfte die früher mehrmals geäußerte Beurteilung der Pflanzensoziologie als geographisch 'im luftleeren Raum' schwebend sich endgültig als ein Mißverstehen erweisen, das allerdings schon lange durch die Synthese von Geographie und Pflanzensoziologie in der Vegetationsgeographie durch Josef Schmithüsen berichtigt worden ist.

8. Planung. Alle großräumigen Planungen, die auf die Pflanzendecke Rücksicht zu nehmen haben oder auf ihr beruhen, werden in immer stärkerem Maß sich den synsoziologischen Forschungsergebnissen bedienen müssen, wenn sie alle Möglichkeiten der Vorausschau ausnutzen wollen.

Schluß. Die Arbeit in der Gesellschafts-Soziologie einer höheren Stufe der einfachen Pflanzensoziologie erfordert allerdings − wie etwa die höhere Mathematik oder andere heutige naturwissenschaftliche Disziplinen (Physik, Chemie, Molekular-Biologie usw.) eine umfassende Kenntnis der Einheiten (Taxa, Syntaxa, Geosyntaxa) über ihre einfachen bis zu ihren höchsten syntaxonomischen Stufen, sowie der Methoden ihrer Erkennung, Darstellung und Anwendung. Eine vielseitige Einführung in diesen Bereich findet sich im Bericht über das 21. Internationale Symposion der Internationalen Vereinigung für Vegetationskunde 1977 in Rinteln (Tüxen, R. 1978).

Dank. Für viele klärende Gespräche über die hier vorgelegten Aufgaben habe ich Prof. Jean-Marie Géhu, Lille, Prof. Karl-Heinz Hülbusch, Kassel, Prof. Rüdiger Knapp, Gießen, Prof. Akira Miyawaki, Yokohama, Dr. Tatsuyuki Ohba, Yokohama, Prof. Ernst Preising, Hannover, Dr. Jes Tüxen, Hannover, Frau Prof. Otti Wilmanns, Freiburg und für mannigfache sachliche Hilfe Herrn Yukito Nakamura, Yokohama (z.Zt. Todenmann) herzlich zu danken.

Zusammenfassung

Nach einer kurzen Darstellung älterer Ansätze und Vorläufer der Synsoziologie, d.h. der Lehre von der Vergesellschaftung der Syntaxa (Pflanzengesellschaften, Biozönosen) werden deren Grundbegriffe kurz dargestellt. In der Reihe der Geosyntaxa, d.h. der Vergesellschaftung der Syntaxa (Knapp 1975) ist das Sigmetum die unterste geosyntaxonomische Einheit, die noch durch Kenn-Syntaxa ausgezeichnet ist. Nach der Entstehung und der Abhängigkeit von exogenen Faktoren werden drei durch Übergänge verbundene Gruppen (Primär-, Sekundär- und Tertiär-Sigmeten unterschieden, deren letzte Gruppe noch weiter aufgeteilt werden kann.

Die Geosyntaxa können nach den Mustern der Sippen- und Synsystematik zu einem abstrakten System geordnet werden, dessen nächst höhere Ränge durch Kenn-Geosyntaxa der nächst tieferen Stufe gekennzeichnet werden.

Die Geosyntaxa können auch nach ihren räumlichen Kontakten zu flächenhaft verbreiteten Komplexen bestimmter Geosyntaxa-Kombinationen gegliedert werden, deren höhere und höchste Einheiten steigende Eigenart (Singularität) besitzen. Das so aufzubauende System (Abb.) ist für die Geographie von vielseitiger Bedeutung.

Auf die verschiedenen Möglichkeiten der Anwendung synsoziologischer Ergebnisse in der Pflanzengeographie, der Vegetationskartierung, der Tiersoziologie, der Pedologie, Geologie und Hydrologie, der Landschaftsökologie, dem Naturschutz, der Geographie und der Planung wird kurz hingewiesen.

Literatur

Braun-Blanquet, J. 1964. Pflanzensoziologie. 3. Aufl.-Wien, New York. 865 pp.

Braun-Blanquet, J. & Tüxen, R. 1952. Irische Pflanzengesellschaften. – Veröff. Geobot. Inst. Rübel, Zürich 25: 224-415. Bern.

Gams, H. 1918. Prinzipienfragen der Vegetationsforschung. – *Naturforschende Ges.* 63: 293-403. Zürich.

Friedel, H. 1956. Die alpine Vegetation des obersten Mölltales (Hohe Tauern). – *Wiss. Alpenvereinshefte* 16: 1-153. Innsbruck.

Hamaya, H. 1975. Landscapes of Japan I, II. Tokyo.

Knapp, R. 1948. Einführung in die Pflanzensoziologie 1. Arbeitsmethoden der Pflanzensoziologie und die Eigenschaften der Pflanzengesellschaften. – Stuttgart. 100 pp. 2. Aufl. 1958. 112 pp.

Knapp, R. 1971. Einführung in die Pflanzensoziologie. – 3. Aufl. Stuttgart 388 pp.

Knapp, R. 1975. Zur Methode der Untersuchung von Gesellschaftskomplexen mit Beispielen aus Hessen und Afrika. – Phytocoenologia 2(3/4): 401-416. Stuttgart-Lehre.

Krause, W. 1950. Über Vegetationskarten als Hilfsmittel kausal-analytischer Untersuchung der Pflanzendecke. – *Planta* 38: 296-323. Berlin.

Krause, W. 1952. Das Mosaik der Pflanzengesellschaften und seine Bedeutung für die Vegetationskunde. – *Planta* 41: 240-289. Berlin.

Krause, W. 1955. Pflanzensoziologische Luftbildauswertung. – Angewandte Pflanzensoziologie 10: 38-40. Stolzenau/Weser.

Der Niedersächsische Minister für Ernährung, Landwirtschaft und Forsten, Hannover. (o. J.) (1978²) Karte der potentiell natürlichen Pflanzendecke Niedersachsens. – Hannover.

Rivas Martínez, S. 1976. Sinfitosociologia una nueva metodologia para el estudio del paisaje vegetal. – *Anal. Inst. Bot. Cavanilles* 33: 179-188. Madrid.

Schmithüsen, J. 1959. Allgemeine Vegetationsgeographie. – Berlin. 261 pp. 1961. 2. Aufl. 262 pp. 1968. 3. Aufl. 463 pp.

Schwabe, G.H. 1977. Ökosystem. Eine Diskussionsbemerkung. – In: Tüxen, R. (Edit.) Vegetation und Fauna. Ber. Intern. Sympos. Rinteln 1976: 555-563. Vaduz.

Tüxen, R. 1931. Die Pflanzendecke zwischen Hildesheimer Wald und Ith in ihren Beziehungen zu Klima, Boden und Mensch. – In: Barner, W. (Hrsgb.) Unsere Heimat: 55-131. Hildesheim.

Tüxen, R. 1934. Vegetationskarte von Niedersachsen 1: 800 000. – In: Brüning, K. Atlas Niedersachsen. Bl. 13. Oldenburg i.O.

Tüxen, R. 1939a. Pflanzendecke. – In: Schnath, G. Geschichtlicher Handatlas Niedersachsens. p. 1/2, Karte 1 b. Berlin.

Tüxen, R. 1939b. Die Pflanzendecke Nordwestdeutschlands in ihren Beziehungen zu Klima, Gesteinen, Böden und Mensch. – *Deutsche geogr. Bl.* 42: 1-8. Bremen.

Tüxen, R. 1942. Die wichtigsten Pflanzengesellschaften der Umgebung Hannover. *Jb. Geogr. Ges. Hannover* 1940 u. 1941: 111-121. Hannover.

Tüxen, R. 1951. Eindrücke während der pflanzengeographischen Exkursion durch Süd-Schweden. – Vegetation 3(3): 149-172. Den Haag.

Tüxen, R. 1956. Die heutige potentielle natürliche Vegetation als Gegenstand der Vegetationskartierung. – Angew. Pflanzensoziologie 13.: 1-55. Stolzenau/Weser.

Tüxen, R. 1957. Dsgl. Ber. z. dt. Landeskunde 19(2): 200-245. Remagen.

Tüxen, R. 1961. Dzisiejsza potencjalna róslinność naturalna jako przedrniot kartografii róslinności. – Polska Akad. Nauk. Inst. Geogr. Przegląd zagranicnej literatury geograficznej 4. Problemy współczesnej biogeografii II: 1-30. Warszawa.

Tüxen, R. 1965. Wesenszüge der Biozönose. – In: Tüxen, R. (Edit.) Biosoziologie. Ber. Intern. Sympos. Stolzenau/Weser 1960: 10-13. Den Haag.

Tüxen, R. 1973. Vorschlag zur Aufnahme von Gesellschaftskomplexen in potentiell natürlichen Vegetationsgebieten. – *Acta Bot. Acad. Sci. Hungar. 19 (14): 379-384. Budapest.*

Tüxen, R. 1977. Bibliographie Symphytosociologica. – Excerpta Botanica B. Sociologica 17(1): 45-49. Stuttgart.

Tüxen, R. 1978. Die heutige potentiell natürliche Vegetation als Gegenstand der Vegetationskartierung. – In: Lauer, W. u. Klink, H.-J. (Edit.) Pflanzengeographie. Wissenschaftliche Buchgesellschaft Darmstadt.

Tüxen, R. (Edit.) 1978. Assoziationskomplexe (Sigmeten) und ihre praktische Anwendung. – Ber. Intern. Sympos. Rinteln 1977. 535 pp. Vaduz.

Wilmanns, Otti & Tüxen, R. 1978. Sigmassoziationen des Kaiserstühler Rebgeländes vor oder nach Großflurbereinigungen. In: Tüxen, R. (Hrsgb.) Assoziationskomplexe (Sigmeten). – Ber. Intern. Sympos. 1977 Rinteln: 287-302. Vaduz.

Zoller, H., Béguin, C. & Hegg, O. 1978. Synsoziogramme und Geosigmeta des submediterranen Trockenwaldes in der Schweiz. – In: Tüxen, R. (Hrsgb.) Assoziationskomplexe (Sigmeten). – Ber. Intern. Sympos. Rinteln 1977: 117-150. Vaduz.

Address of the Author:
Prof. Dr. Drs. Lit. Reinhold Tüxen
Arbeitsstelle für Theoretische und Angewandte Pflanzensoziologie
Todenmann

THE HISTORICAL ECOLOGY OF THE ROYAL FOREST OF WOLMER, HAMPSHIRE

E.M. Yates

'Nos saltus viridesque plages camposque patentes
Scrutamur totisque citi discurrimus arvis
Et varias cupimus facili cane sumere praedas'

Nemesianus

Introduction

Royal forests in England were districts subject in medieval times to a special body of law, forest law, designed to maintain game for the King's hunting. The 'beasts of the forest', as they were described, included the red deer, boar and wolf but most of the preservation embodied in the law related to the red deer. The areas to which the laws applied contained a variety of land with wood, heath, pasture and arable in varied proportions, but through which the deer were free to roam. Some forest land was held directly by the crown, but some was held by other feudal magnates or by their free tenants.

Despite the intermixture of lands of various types the royal forests generally contained tracts of uncultivated land, some heath and swamp, some remnants of primaeval woodland. The survival of these unimproved lands through previous periods of woodland clearance was probably due to difficulties of the soil, either too sterile or too heavy. The distribution of royal forests is therefore related to geology. There is a marked correlation for example with the Bunter Sandstone outcrops, as with the Forest of Morfe and Sherwood Forest, and with Cretaceous and Eocene sandy formations (Aptian and Lutétien) as with Wolmer Forest, Windsor Forest and the New Forest.

The laws, enforced most strongly in the early medieval period, had the effect of delaying clearance for agriculture of the woodlands to which they applied, and were significant in this respect as land hunger increased with the increase in population. After the mid-14th century the laws were less vigorously enforced, and some woodlands were disafforested. By then epidemics had reduced the population pressure with the result that large areas of woodland survived into modern times, in some instances until the present day. Interest of the crown in the royal forest was fitful after the 14th century and their administration decayed. From the 18th century they were seen either as of value for scientific forestry, particularly with regard to the needs of the Navy, or as land for clearance and cultivation. Today the royal forests, or fragments of former royal forests, are important as nature reserves, and as amenity areas. Because of this combination of historical and geographical causes they contain remnants of primaeval oak woodland, much altered during a chequered existence, but nevertheless continuously forest since the re-establishment of woodland after the Ice Age, that is to say for upwards of 8000 years. On the European continent the earlier establishment of scientific forestry, and the lesser power of the central authorities in medieval times, has tended to militate against such survivals but the woodland of Bialowicz in east Poland owes its survival to its status as a hunting forest of the Russian tsars. Similarly part of the Forstenrieder Park south of Munich owes its survival as woodland to its use for hunting by the Wittelsbachs.

Forest Law and its administration

Forest law, it is widely believed, was imposed by the Norman Kings, but may well have its roots in previous centuries. Saxon England was hierarchical in organisation and it is most unlikely that hunting rights were equally distributed to all levels of society. The laws were an aspect of the power struggle between the King and his nobles. The King claimed the sole right to hunt over the forest lands, including lands held of the crown by the feudal magnates. When the crown was weak and or impoverished the laws were lessened in severity and the area to which they applied reduced. In William I's reign the forests were extended (H. Arthur Doubleday & W. Page 1903). Henry II in the Assize of Woodstock 1184, the earliest authoritative document dealing with forest law, set out the limitations on land usage imposed by the crown. These included the prohibition of felling by land owners except for fuel, and even then only in the view of the forester. In John's reign the forest boundaries were extended but during the minority of Henry IIIs, the charter of the forest was issued on November 6, 1217 in the King's name. This, and the perambulations of the forest boundaries in the following year, removed from the forest all lands afforested by Richard I and John, and much of the land afforested by Henry II. The charter of the forest marks therefore the end of the expansion of the forests and the beginning of a long period of attrition. It also restricted the number of attachment courts, where offences were first considered before reference to itinerant justices, to three a year, and permitted fencing-off arable lands not held directly by the crown. Furthermore the charter promised that no man henceforth should lose life or limb for poaching. (J. Charles Cox 1905). The charter was reissued by Edward I in 1299 and in 1301 large areas were disafforested, as part of the financial preparations for French and Scottish wars. Disafforestation could be bought from the crown. In 1190 the Knights of Surrey bought disafforestation of the south of the county for 200 marks.

The administration of the forest on the judicial side was primarily the function of the forest justices, and their hearings, the forest eyres, were to be held every seven years, although in practice the period varied. The hearings took place at a town adjacent to the forest. The justices with their servants and clerks, the various offenders with their sureties, and the foresters, must together have made for considerable occasion in the life of small market towns. The circuits followed by the justices were long. In 1335 John of Londham on circuit was first at Guildford to deal with offences relating to forest of north-west Surrey, then at Southwick for hearings relating to the Forest of Bere juxta Portchester, at Lindhurst for the New Forest, at Ashley for the Forest of Bere juxta Winchester, Salisbury for Clarendon Forest and Hockholt, Chippenham for Melksham Forest, and then into Dorset.(1) The executive side of the administration and the initial stages in the judicial procedure were in the hands of the chief forester of the crown, but usually supervised at a local level by a warden or bailiff. The actual administration was performed by four verderers, elected by the freeholders. The day-to-day patrolling, and arrest if necessary, was performed by foresters. Other officers were the agisters who supervised the permitted grazing of the forest lands by stock, and collected payment of same. The grazing was shared by people of the peripheral parishes as well as parishes within the forest, and was prohibited during the midsummer period to prevent disturbance when the deer were breeding, and in midwinter when feed was short. There were also woodwards responsible for the timber, and finally the regarders. The latter were usually local people of some standing who at the regard had various questions to answer about the state of the forest and its boundaries. These questions covered twelve separate matters of interest to the crown such as pasture within the forest, clearance and fencing-off land, presence of hawks, honey and the keeping of dogs. Dogs kept within the forest were to be hambled or lamed by mutilations of the ball of the hind foot, unless their owner paid for the privilege of keeping dogs not so lamed. Regards were held in theory every three years. (G.J. Turner 1899).

The local courts at which offenders were initially charged, and at which they provided sureties for their attendance at the next forest eyre, were the courts of attachment or swainmote, held within the forest several times a year, but after 1217 limited to 3 times a year. The penalties at the forest eyre were severe, but varied according to the social status of the offender. For low-born offenders at different times previous to 1217 the penalties for hunting had been blinding, amputation of the right hand, skinning, and hanging. It is a curious social commentary that the death penalty for poaching accompanied by destruction of woods, was re-introduced in 1722, without benefit of clergy. This refers to the right of men in holy orders to be tried before ecclesiastical courts and not by criminal courts. Benefit of clergy was a very considerable privilege that at the higher level played a part in the struggle between Becket and Henry II. At the lower level it had the effect of including many churchmen in the list of poachers reported at the forest eyres.

Extent of Wolmer Forest in the 13th century

The name Wolmer Forest (orginally Wolfmere, now often mispelt Woolmer) applies now to an area of heath and coniferous plantation east of Liphook on the main London-Portsmouth road the A3. North of Wolmer Forest. Separated

from it by a belt of farmland is an area of oak and conifer plantation administered by the Forestry Commission and known as Alice Holt Forest. In medieval
times these two areas were administered together and Alice Holt was often regarded as part of Wolmer Forest. In 1278 for example the coppice known as the
Strait in Alice Holt was described as in Wolmer Forest.(2) In 1304 a list of wastes
made within Wolmer Forest included the Frith, also in Alice Holt.(3) In 1327
Alice Holt is named as within Wolmer Forest (4) (Fig. 1). Conversely in the Pipe
Rolls for 1190 William de Venuz is recorded as the warden of Alice Holt Forest,
and Wolmer is included.*(5)

The pleas of the forest for 1270, held at Winchester, provide the earliest
known boundaries for Wolmer Forest (6) and are repeated in a Chancery document of the same period.(7) The area subject to forest law was then much greater
than the present extent of both Wolmer Forest south and Alice Holt. It extended
from Wrecclesham parish boundary in the north to Sheet Bridge in the south and
from the Surrey boundary westward to Alton. (Fig. 2 and 3). It also extended
southeast into Sussex. Much of the outcrop of the Hythe Beds is poor land, still
much wooded (Figs 1 and 2). Harting Combe in Rogate Parish, and within this
belt, is described as within the forest in 1281(8) and although by 1367 the lands
of the Abbot of Durford were outside the forest he was assessed for a 'paturam'
(the payment by landholders to foresters) of nine loaves, six gallons of ale and
three salt fish (ling or conger).(9) In 1270 perambulations named Chithurst and
Sheet Bridge so that the south-east boundary was near the River Rother. Indeed
the place name Rogate is probably a reference to this location of the forest
boundaries.

The 1270 forest therefore included a range of soils, varying in quality from the
heavy clays of Alice Holt on the Gault and the sterile sands of Wolmer Forest
south on the Folkestone Beds, to the high arable loams around Selborne on
the Upper Greensand and Lower Chalk plateau (Fig. 2). As is to be anticipated
these high grade soils were the first to escape from the forest law, excluded after
1270. The widening of the outcrop of the Gault and the Folkestone Beds in this
part of the Weald is due largely to structural causes. It is this variation in geological structure, with its resulting wide outcrop of poor or difficult soils that is part
of the explanation for the original existence of Wolmer Forest. Nevertheless outcrops of the Folkestone Beds in the Weald are in general little cultivated so that
the survival of uncultivated land on the Folkestone Beds within Wolmer Forest
cannot be attributed solely to the forest laws. It is likely however, that in contrast,
the survival of the Alice Holt woodlands can be so explained since similar clay
soils elsewhere in the Weald were taken into cultivation in the 13th century. The
laws have helped to preserve also other scraps of woodland outside the present
boundaries. The combination of physical causes and historical causes, particularly
the forest laws, gives within the 1270 boundary four sub-regions or Landschaften

---

* In this paper 'Wolmer Forest' will be used for the whole of the ancient forest, and Wolmer
Forest south for the present-day area.

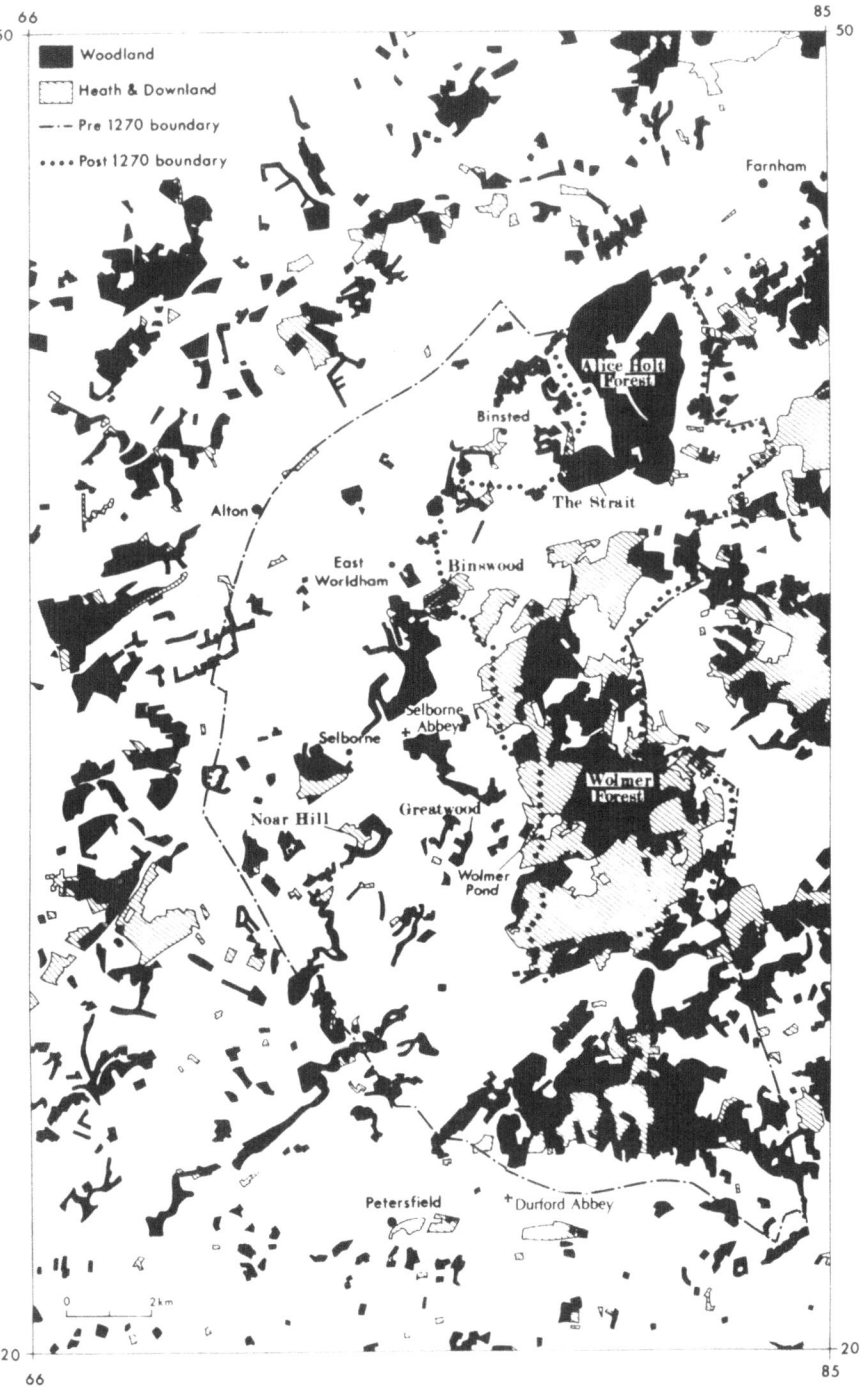

Fig. 1 The land-use of Wolmer Forest. Based on L.U.S. maps of 1931-2.

97

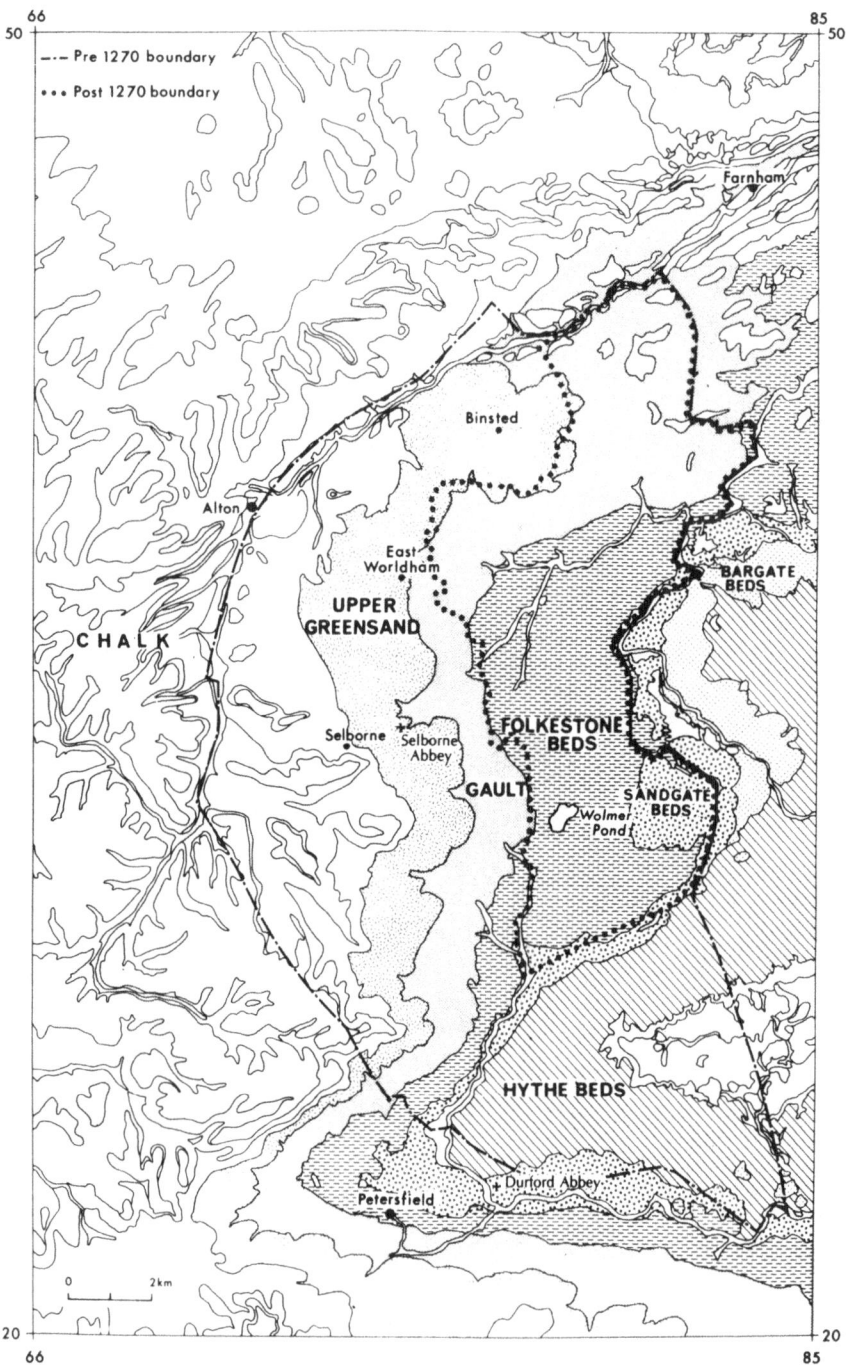

Fig. 2 The geology of Wolmer Forest

98

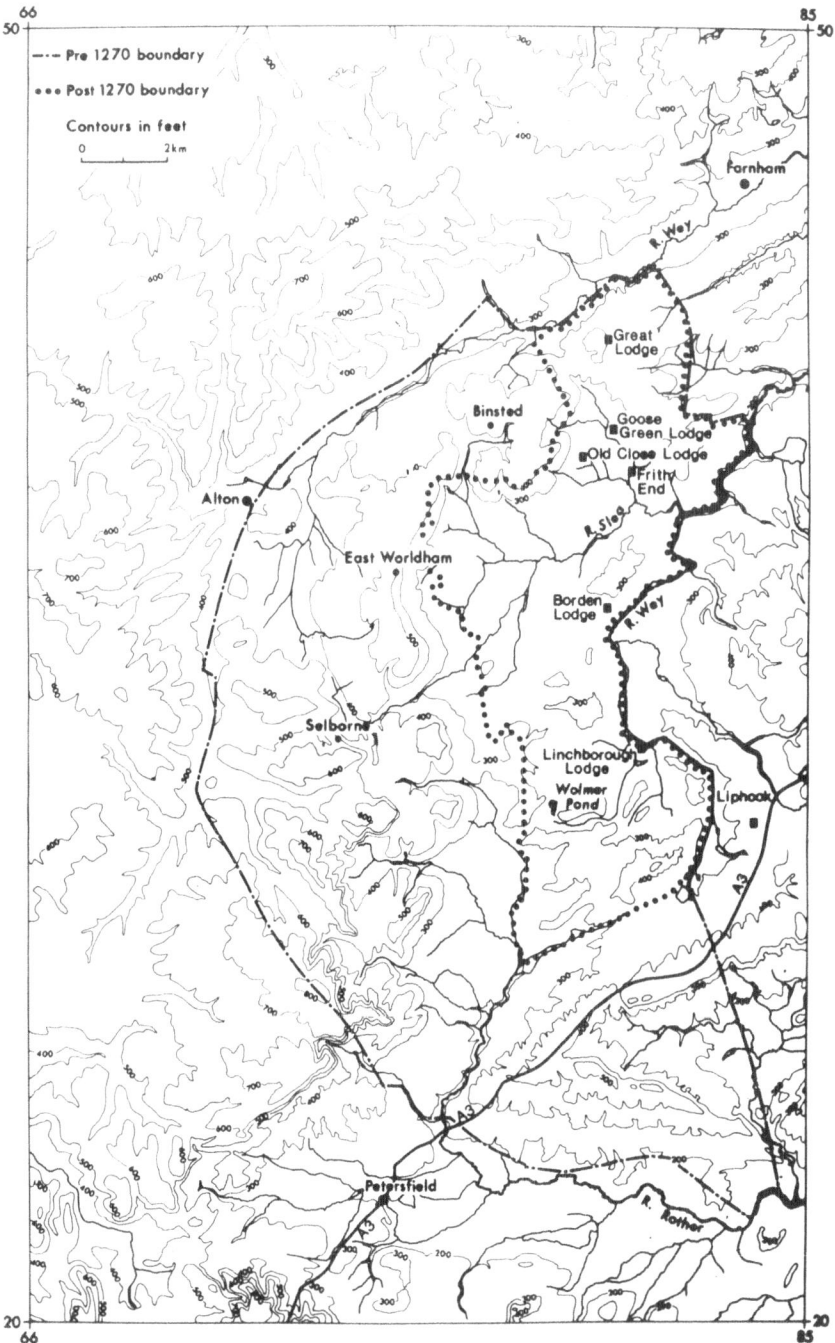

Fig. 3 The relief of Wolmer Forest

(fig. 1) the arable plateau extending around Selborne, East Worldham and Binsted with isolated copses and small woodlands mainly on steep slopes, associated with the outcrops of Upper Greensand and Lower Chalk; the oak and conifer woodland of Alice Holt on the Gault; the heath and woodland of Wolmer Forest south on the Folkestone Beds; the conifer plantations and sub-spontaneous woodland on the Hythe Beds in north-west Sussex.

It is difficult to say whether the 1270 boundaries represent the maximum extent of the forest. A large part of Hampshire was under forest law and it was possible to travel from Windsor to the New Forest, entering Hampshire at Eversley, without leaving the area subject to forest jurisdiction (D.M. Stenton 1951). Possibly Wolmer once extended further west and linked with the tract of forest through Eversley and Pamber. A large part of Hampshire was disafforested in 1225 after the charter of the forest of 1217.(10) The land concerned had been afforested between the coronation of Henry II and the ninth year of Henry III. By this afforestation, revoked in 1225, the royal forest of Windsor had extended across the county boundary to Basingstoke and Odiam. On the Surrey side, forest reached almost to Guildford and Farnham so that the break from Wolmer Forest was a narrow strip around Farnham.(11) At an earlier date there was no break, for in a forest eyre of 1345 the men of Farnham, Wrecclesham and Headley were accused of grazing their stock in Wolmer Forest despite the fact that their vills, all held by the Bishop of Winchester, had been disafforested. (12) This disafforestation is a further example of the tendency for regions with better soils to be removed from the forest. Much of Headley is situated on the outcrop of the Sandgate and Bargate Beds, between the Folkestone Beds and the Hythe Beds, and giving rise to first-class soils.

The part of Wolmer Forest in Sussex had also been more extensive. Even in the late 13th and in the 14th century the forest eyres still record poaching offences east of the pre-1270 boundary, usually with the deer allowed to enter parks and chases by a deer leap but being unable to return. The park at Verdley north of Midhurst is mentioned in this context(13) and the park of the de la Zouche family at Treve (now River) near Petworth, even further west.(14) This form of poaching also applied in Surrey for the men of Lodsworth (Sussex) and Chiddingfold (Surrey) together with the servants of the Prior of Shulbrede (near Haslemere) were similarly charged with constructing a deer leap into the park of the Bishop of Winchester. (15) The same device, a deer leap, was also employed on the other side of the forest at Le Court in Greatham.

The Wolmer countryside in 1270

The boundaries of 1270 included many settlements and therefore large areas of arable land. Traces of open field agriculture survived at Selborne into the 19th century, and open field was present at other settlements on the Upper Greensand such as East and West Worldham (G.I. Meirion Jones 1968). Some openfield was also present, more surprisingly, in the regions of poor soil, as at Liss(16) and Greatham(17), but generally on the heavy clays and poor sands extensive woodland was present, some surviving until modern times. Even on the Upper Greensand, Lower Chalk Landschaft, woodland was present in the Dark Ages, in some instances surviving into medieval times. The cleared land of Selborne, Farringdon, Hawkley etc. in the Dark Ages must have formed islands in the woodland whereas today on the Upper Greensand, woodland is infrequent, patches surrounded by farmland. Evidence for this former woodland is contained within the place-names of the Upper Greensand, Lower Chalk Landschaft. The name 'Rhode Farm' on the edge of the Upper Greensand escarpment in Selborne is derived from the verb ridden = to clear ground of trees. Hartley Maudit is 'the clearing of the hart' (plus the personal name Maudit), Bentley the clearing where the bent grass grows, and Wheatley the wheat clearing. Many of the places peripheral to the forest have the place-name element scéat, becoming shott or sheet in modern English. This element, meaning a corner, occurs not only on the poorer soils such as Bramshott (bramble + scéat) and Sheet, but also at sites on the Lower Chalk and on the Upper Greensand such as Oakshott, Empshott and Bradshott (Empshott = Imbes céat, the shott with the swarm of bees, Bradshott = the broad shott). These shott names were probably associated with projecting corners of an extensive area of woodland at its outer margins. The curious pattern of hundreds with various detached areas is also argument for the former extent of woodland (Fig. 4).

Some of the actual work of clearance can be dated more precisely in the 13th century by reference to the records of the forest eyres in which details are given of purpresture and assarts on royal and other lands.(18) Some of the assarts related to land on the Upper Greensand and Lower Chalk, some to the more difficult land of the Gault and Folkestone Beds. A few examples can be quoted. Godfrey of the Fen had permission to assart in Newton on the land held by William de Valence (the Valence family held Newton between 1249 and 1323, giving their name to the villiage). (W. Page 1908) William de Valence also held Hawkley, for long part of Newton Valence parish, and Elias of the Mill was granted permission to assart in Hawkley. An assart by Ralph le Farway, also in Hawkley, was in Snellinche (now Snailing) and probably on the Gault. Thomas Thurstayn asserted on lands of the Prior to Southwick in Dean (now Priorsdean), and further assarts in Dean were made by Peter le Bel of Petersfield. Henry le Fuller of Headley had assarted at Iverley (now Eveley) on the sands. These additions to the cultivated land, transforming a predominantly wooded landscape to one of isolated woods and copses set in fields, were widespread in the 13th century forest eyres, and as shown it is possible to identify not only the areas and the landholders, but, lower down the social scale, the people involved in the actual manual work of clearance, perhaps in some instances those wielding the axe.

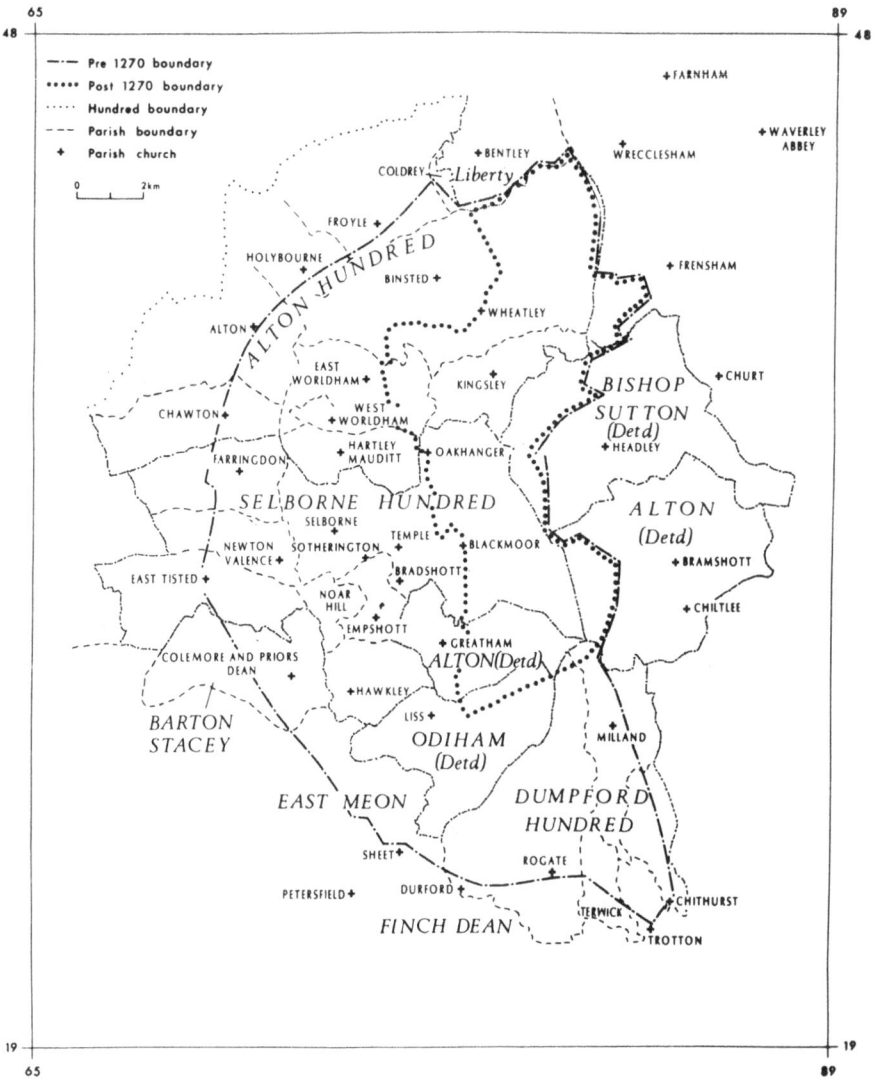

Fig. 4 Parishes and hundreds of Wolmer Forest

The woodlands

The later references to surviving woodland show that it then consisted predominantly of oak and beech in a ratio of three to one, plus small numbers of ash and a little elm (Commissioners of woods and forests, 1790). The references in the forest eyres to felling in the forest also show a predominance of oak and the beech in the 13th century. In 1270 the unlawful felling of 30 beech and 35 oaks was reported.(19) Some of the offenders came from outside the forest and the site of the felling cannot be identified, as for example John of Lasham who had felled ten oaks. In other cases the sites can be identified, for example the two oaks felled by Gervase of Snellinche (Snailing), the single oak felled by Gerard of Wyck and the two oaks felled by William de Nevill of Binsted. It might be anticipated in view of this evidence and general considerations of site that the beech was more present on the drier soil in the Wolmer Forest south Landschaft, but the 1790 report makes it quite clear that beech was also presort on the heavy clay soils. A survey by the regarders in 1576 noted sale of beech timber from the sandy areas whilst the 1790 report also noted well grown oaks in the Linchborough Lodge enclosure (Fig. 3) on very poor sands. The evidence then is for an oak/beech woodland in the eastern Landschaften. Forming the lower storey in the woodland were maple, holly, blackthorn, hawthorn, crab with also willow and alder. References to maple occur in the 1576 regarders' surveys, and blackthorn in 14th century eyres.(20) Holly was specifically mentioned in an Elizabethan survey of destruction and enclosure by Lord de la Warre as will be subsequently discussed. (21) The latter, originally of the West family of Oakhanger, had become warden in 1579, and rebuilt Great Lodge (Fig. 3) with some felling for constructional timber. Similar destruction had taken place under the previous warden Thomas Paulet, deputy to John, Marquis of Winchester. The Paulet family held Norton in Selborne. During the long enquiry with numerous witnesses the pressure on the woodland was clearly shown and also something of its composition. Thomas Acoste of Bentley for example had bought hollies and thornes 'those were such as a bucke might turne over with his hedde.' John Harper of Bensted had used hollies and thornes for levers and rollers to extract oak, similarly Humphrey Tanner of Frensham. Holly was mentioned in the 1790 report; Hollyhill is a place-name near Linchborough, and some large hollies still survive within the forest. In all the scattered strands of evidence show an extensive mixed oak/beech forest with a well developed lower storey. The last piece of evidence to be mentioned is also one of the earliest. As previously mentioned, in 1278 instructions were given to Edmund Thurston and Peter atte Marsh to make a 30 acre coppice in the Strait in Wolmer Forest, excluding game and domestic stock. When the coppice was cut, oak, ash, beech and crab were to be left.(22)

It is of course unlikely that the woodland extended across the Gault and Folkestone Beds' Landschaften without break. The placename Headley (= heather-covered clearing) is evidence that heath was present in the Dark Ages, as might be anticipated from the various tumuli showing prehistoric occupance. It is equally obvious that since one area was distinguished as a heathy clearing, the district was not otherwise heath. The earliest evidence as discussed above suggests that woodland was extensive on the Folkestone Beds and, when effectively protected, survived to the 16th century. In the plea rolls of the 14th century there are entries recording unlawful burning of heath. In 1345 Thomas Fry of Langley (in Rogate) burnt 100 acres.(23) Fry was described as a common trouble-maker, often poaching with dogs and nets, so that the burning may not have been for husbandry. On 1st April 1372 however one hundred acres were burnt at 'Severeshull' and 'Weverston' (= Weavers Down) by William of Chiltlee and John le Monk his servant. On 20th April 1372 Richard son of Thomas le Weyn of Greatham burnt 15 acres at Whitepits.(24) This spring burning was probably to ensure a good growth of forage. Burning plus turf cutting, and the grazing pressure to be discussed, would explain the extension of the heath until Wolmer became the treeless expanse described by Gilbert White (G. White 1789). The Dark Age landscape of the Folkestone Beds was probably wooded with patches of heath adjacent to the various meres formerly present. Some of these meres such as Blackmere (= Blackmoor) Hogsmere, Tangmere and perhaps even Longmere (= Longmoor) have gone but the 'Wolves mere' and Cranmere survive with reduced extent. The hardpan responsible for the meres led also to peat growth, giving areas of acid bog and wet heath, as well as dry heath on the more elevated or sloping sites.

The fauna of the forest is still of interest in that the natterjack toad (*Bufo calamita*) and smooth snake (*Coronelle austriaca*) survive. Deer are again present, and the surviving meres are important for water fowl. Indeed the avifauna is surprisingly rich, due paraxodically to the present use of much of Wolmer Forest south for army training, with limited public access.

Once however the whole area must have been a paradise for wild life, and although much survives, much has gone. The wild boar that gave its name to Borden in Chithurst, and to Borden in Headley (now most curiously spelt Bordon, making etymological nonsense), has gone. The wolf that gave its name to the mere and thence to whole forest has also vanished although surviving at least to the late 13th century. Adam de Gurdon was granted free chase for wolves and hares in East Tisted in 1275.(25) The crane of Cranmere has not bred in Britain for centuries. The 1270 forest eyre records two other features of the wild life now much reduced: the wild bees (the honey belonged to the crown) and the falcons, likewise a crown possession.

The use of the woodlands

The major function of the forested lands was the provision of game for the King. A royal hunting lodge provided accommodation. and the royal party also stayed from time to time in Durford Abbey, Waverley Abbey and Selborne Priory. Edward I's movements are fully known. He stayed in the lodge in Wolmer Forest August 1-7 1285; April 25, 1290; August 28-30, 1299; and August 28-30, 1302. He stayed at Selborne Priory in August 1276 and July 1280, Durford Abbey in 1276 and 1294, and at Waverley Abbey 1292, 1299 and 1303.(26) The site of the lodge has not been identified. There are some suggestions that it was at Lode Farm near Kingsley. Certainly there was a royal chapel in Kingsley, recorded in a dispute involving tithes (27) and it is likely that the lodge would be in the vicinity. Capes suggests Bramshott as the site of the lodge, built in 1285 with a garden for the queen (W.W. Capes 1901). This suggestion is based on the fact that several royal documents were issued from Bramshott. The presence of a garden would imply more than a simple lodge and this is supported by a record of payment made in 1285 for repairs over a period to saddles and harness kept in Wolmer Forest.(28) Gilbert White mentions East Worldham in this context, for long associated with the wardenship of the forest, and where there is still a King John's Hill (G. White 1789). On the other hand the lodge was repaired in Edward III's reign with 4000 bricks and 130 ridge tiles bought of John Walter of Flexham (in Liss), suggesting a more southerly site.(29) Further repairs were made in Richard II's reign.(30) Folk memory in the area also associates Noar Hill with a King John hunting lodge.

Visits to the abbey and priories by royal parties were very expensive for the hosts. After his stay at Selborne Priory in 1286 Edward I gave the prior six oaks in recompense, a use of the forest to pay debts (W. Page 1908). Edward II gave Durford Abbey £10 after a stay there, a very large sum of money. (W.W. Capes 1901).

The hunting was of red deer and fallow *Cervus elaphus* and *Cervus dama*, both species being recorded from all parts of the forest in the forest eyres despite their latter separation into red deer in Wolmer Forest south and fallow in Alice Holt. (Rogate' is evidence that roe *Cervus capreolus* was also present. It was poached even if. regarded as inferior for sporting purposes. Wild boar survived in Sussex until the 12th century but apparently does not appear in the Wolmer Forest records. The hunting was with dogs of which two types are named, one resembling a hound, the other a mastiff, They were also used by the poachers. In 1329 for example Thomas de Burhunte (Bohunt in Bramshott) hunted illegally with three hounds ('Leporar') one red, one white, one brindle.(31) The bow was used and presumably the spear. In poaching, as discussed below, traps and nets were employed. The nobility hunted on horse, so that provision for horses and dogs had to be made. In 1314 stumps were taken from Wolmer Forest to provide heating in the kennels, and in 1315 similarly to warm the stables of the King's horses at Odiam (W. Page 1908). According to Capes the Lord of the manor of Oakhanger had to provide a pack of white hounds when the King hunted (W.W. Capes 1901). An amusing aspect of the hunting parties is provided by Gilbert White. Morris Ken the jester fell from his horse so many times to Edward II's amusement that he was rewarded with part of Wolmer Forest Lodge (G. White 1789).

The monasteries and priories may have suffered from the need to provide accomodation for the King but they certainly provided, together with the upper levels of society, many of the poachers, either because of clerical immunity or because of church ownership of lands adjoining to Wolmer Forest. Examples are numerous. In 1372 Richard le Walshe servant of the Prior of Selborne hunted with three mastiffs at Cranmere, Brothers Thomas and Edward of Selborne Priory hunted with six hounds, and Brother Edward killed a deer in Hartley Wood.(32)

In 1377 a group of poachers included the Abbot of Durford, John Ryhull his chaplain, Henry Husee Lord of Harting, John Bull of Greatham, and Robert Marcant of Mapledurham.(33) Examples of secular clergy involved in poaching were Robert parson of Headley(34) and John Morlyn parson of Hascombe.(35) As noted, poaching was often with nets. Nets were used in the early 14th century by various servants of the Bishop of Winchester at Crondall, on the northern boundary of the forest.(36) In 1333 in hearings before John of Macclesfield, nets and other traps were reported as used between Pamber Forest and Windsor Forest so the practice was probably widespread.(37) The poaching provided from the forest an important source of protein as well as sport. The crown made presents of venison, and venison was supplied by the foresters to the court. The forests were therefore a source of meat, although the deer required some supplementary feed in winter.

Another major use of the woodland and waste was grazing. Those living within the forest and in peripheral villages had the right to graze their cattle except in the closed month, and when feed was short. According to keeper John Adams, in the 6th report, cattle were branded with a forest mark and marked on the ear for individual ownership. The cattle were driven in fence month into Wolmer Pound, and cattle grazing without right were separated. The grazing was for cattle and pigs. The pigs were thought to prevent too much mast being eaten by the deer, with poisonous effect. By Gilbert White's time, geese were grazed in the forest in large numbers. Payments made for grazing were recorded at the forest eyres and in their fluctuations give an indication of weather conditions in the 13th century (1247-48 to 1270-71, 28s-0d, £17-16s-$3\frac{1}{2}$d, 24s-4d, £8-10-0d, £7-14-$9\frac{1}{2}$d, 12s-0d, £4-11s-$0\frac{1}{2}$d, 27s-0d, 31s-0d, £12-2s-6d, 40s-$4\frac{1}{2}$d, 7s-3d, 3s-8d, 7s-2d, 11s-8d, 3s-3d, 117s-4d, 13s-5d, 14s-9d, 2s-3d, 22s-10d, 18s-2d, 7s-8d, 8s-4d).(38)

The woodlands also provided, as is obvious, timber for building and for fuel. As with hunting, timber extraction went on both illegally and legally. Legal gifts of timber were made by the crown, whilst the justices in eyre dealt with illegal felling. An example in 1361 is John Cambeley who was fined for using good timber to heat an inn within the forest. The dead wood, faggots, top and lop provided fuel but in the 18th century the forest was providing a further source of fuel in peat and turf. The quantities were large. In 1782 for example Borden Lodge produced 325 loads of peat, 340,000 turfs and eight kilns of heath, the last named being used in brick making. There is little evidence apart from a 13th century reference as to how long these other fuels had been used, but obviously it is likely the practice of peat and turf cutting was ancient. In the 18th century the growth of rural population may have increased the pressure on the forest but it is probable that the poor had always been dependent on this form of fuel, contributing further to the devastation of Wolmer Forest south.

Bracken was also a valued product of the forest, used amongst other things for cattle litter. In a 16th century dispute about the location of Blackmore pound the cutting of bracken was said to be a perquisite of the keeper of Borden Lodge.(39) The forest also yielded wild fruits. Naturally there is little literary reference to this but by the 19th century cranberry *Vaccinium oxycoccus* was methodically gathered for sale by the poor.

The woods and heaths of the forest had another significant role; they constituted land for expansion. This was a matter of survival for the peasantry as population grew, and was also important to higher social classes in the search for landed estates. Finally the forest was a refuge for the lawless and the destitute. The road through Alton in the 13th century was noted for the risk of armed robbery, and seven merchants were murdered in Wolmer Forest in 1298 (W.W. Capes 1901). The eyre rolls of 1361 refer to the imprisonment of paupers arrested by foresters within the forest.(40)

Administration in the 13th century

The chief keepers, usually described as baillivus, during the early 13th century were drawn from the de Venuz family who held lands at Empshott, within the forest. There are references to the family as far back as 1130 in the Pipe Rolls, and John de Venuz is named in 1247 on the earliest detailed account taken at an eyre that survives.(41) He was also one of four agisters with Hugh the beadle of Wheatley, Stephen Wileking and William son of Walter of Kingsley, all obviously local men. He also stood surety together with Adam de Gurdon for John of Bramshott and John of Westcot (in Binsted) when they had poached in Wolmer Forest. The villages of Chiltlee, Liss, Liss Abbess, Langley and Harting were held responsible for not raising the hue and cry when this poaching offence was committed. At the forest eyre in Winchester in 1270 it was reported that John de Venuz, son of John de Venuz, was under age and that the King had granted the forest to Adam de Gurdon who had married Constance de Venuz.(42) In 1300 Richard de Westcot and Peter de Heighes (of Blackmore) were verderers. The repetition of family names in the records over long periods show that the offices tended to become heridritary. It also suggests that enforcement of the law must have proved difficult in that it was local landholders who were required to enforce the law, but who stood to lose by its enforcement. It was also this class, together with clergy, that was responsible for much of the poaching.

The forest in the 14th century

After 1270 the forest was reduced in area, and the new boundaries maintained until the 19th century.(43) The forest was restricted to two Landschaft units, the Folkestone Beds and the Gault. For this reduced forest the eyre courts continued to function throughout the 14th century although becoming more perfunctory. In the first half of the century in addition to poaching offences, great detail is given of clearances for agriculture, followed by hedging and ditching. The clearances recorded in the 13th century were continued, but after the middle of the 14th century the references to clearance disappear. This can be exemplified by refence to an enquiry made in 1304 into the clearances.(44) The assarts were listed giving amount and value per acre. Rents were highest at Bradshott and Sotherington on the Upper Greensand, and lowest at Oakhanger on the sands.

Land holder	place	acreage	rent per acre in pence
Peter de Heighes	Blackmore	17	4
Jacob de Norton	Blackmore	13	4
The Templars	Sotherington	22	5
Roger de Bradeshute	Bradshott	$4\frac{1}{2}$	5
Thomas Paynel	Blackmore	10	4
	Oakwood	15	4
Henry Cosyn	Wheatley	5	4
	Kingsley	$1\frac{1}{2}$	4
Richard de Westcot	Kingsley	$6\frac{1}{2}$	4
Peter de Heighes	Conbridge	$7\frac{1}{2}$	4
John de Drokensford	Kingsley	4	4
	Conbridge	$4\frac{1}{2}$	4
	Frith End	1	4
Walter Whiteloke	Frith End	$\frac{1}{2}$	4
Peter le Crokkere	Frith End	$\frac{1}{4}$	4
Stephen le Crokkere	Frith End	1	4
John de Thedden	Isington	2	4
John de Benstede	Binsted	3	3
Thomas Paynel	Oakhanger	15	2

As can be seen clearance was still being undertaken by the landholders such as Jacob de Norton (in Selborne) and Peter de Heighes, and peasants such as Stephen le Crokkere and Walter Whiteloke.

The clearances were widespread, affecting all parts of the forest, but having greatest impact on the Upper Greensand and Lower Chalk plateau Landschaft, finally creating the present-day open scenery with woodlands surviving primarily on the scarps and steep slopes (Figs. 1 and 3). In the 1345 eyre the groves within the forest not in direct possession of the crown were listed (45) including those held by the Popham family and therefore in Farringdon, and that of Jacob atte Overe (Noar Hill). The beech hanger on Noar Hill is therefore a remnant of the original woodland. As previously mentioned in relation to 13th century clearance, the lords of the manors listed in 1304 were of course not directly involved in clearance themselves and the large acreages were probably made up of numerous small plots rented by peasants who undertook the actual clearance. This is shown in the eyre rolls of 1337. The cleared land held by the Templars in Sotherington and listed as purprestures was made up of eight $2\frac{1}{2}$ acre blocks, each with the occupier named, including William Holeway, Stephen atte Lynche, John Stalbat and John le Feghelere. (46)

Many of the assarts are described as old and they show the former extent of the forest. They are listed by township including Kingsley, Binsted, Wheatley, Blackmore, East Worldham, Alton, Norton, Farringdon (held by Richard of Popham from the Bishop of Exeter, as mentioned above), and Noar. Similarly old purprestures included Hawkley, Greatham, and Sotherington. Payment to avoid hambling of dogs is also evidence of the former extent of the forest. The Prior of Selborne and Simon de Heighes (Blackmore) made such payments, and also Rudolph de Camoys who held land in West Tisted. The same landholders also paid for the right to have windfall wood. On the Gault outcrop in the south some extensive patches of woodland survived this period of clearance and have remained woodland until today.

The administration of the forest in the 14th century was once more with the Venuz family, and associated with the manor of East Worldham. As the century progressed twelve regarders were listed at the eyres, including representatives of the families Heighes (Blackmore) atte Rythe (Liss), and le Bel (Petersfield and Priors Dean). Presumably the regarders lands were well distributed in relation to the forest to simplify supervision.(47) Verderers in 1333 were William of Bramshott and William of Rotherfield (East Tisted) from the eastern and western extremities of the forest.(48)

The forest in the late 16th and early 17th century

The long sequence of eyres comes to an end in the 14th century and there is little trace of the administration of the forest until Elizabeth's reign. The wardenship passed to the Burghersh family and then by marriage to William de la Pole, Earl of Suffolk who was described as holding lands formerly held by John de Venuz.(49)

It is well known that the Stuart Kings endeavoured to increase royal revenue by revitalising various feudal practices, including those relating to royal forests. This movement was however already underweigh in Elizabeth's reign, and documents relating to the forest became more numerous from 1570. These include regarders reports showing that though some woodland survived in Alice Holt, little was left within the Wolmer Forest south. The last remnant of woodland was reported as felled by persons unknown in 1578.(50)

The crown's interest in Wolmer Forest is shown by two enquiries. Reference has already been made to one, touching on the right to cut bracken. The other, also mentioned, was the result of a complaint by John Taverner, surveyor general of her majesty's woods.(51) The main defender was Lord de la Warre, who became warden or bailiff in 1579. The enquiry was based on forty-two predetermined questions, put to a series of witnesses in Alice Holt on the 15th April. From the depositions or answers to these questions it emerged that all the lodges were ruinous when de la Warre became warden. The old lodge was then rebuilt into a very fine house, in itself requiring the felling of 120 oaks. Other trees had been felled to repair the sluices to the various meres (used as fish ponds) and to fire the large quantities of bricks required in the building. The distinction between crown property and personal property of the warden had become blurred, and the hunting lodge had become a fine private residence. Other depositions show very large quantities of wood were being sold to clothiers, weavers and dyers in Farnham, and various timber workers, lathe makers, carpenters, colliers and sawyers, in Frensham. Many trees had been pollarded. Even the regarders were involved. Sir John Freland the regarder for the southern part of the forest had bought 'one frame of tymber ready framed fitt for an odehous havynge in it two rooms which was carryed away at viii or ix loades the wayes being so fowle as they would not carry above one tonne at a loade'.

The enquiry also showed that eleven new cottages had been built on forest lands. Altogether some 271 great trees had been felled during the early part of de la Warre's wardenship, including many that had been marked with 'three scotches' by Mr. Taverner at a previous inspection. The royal forest had reached its nadir. The wardens saw it as a private resource, settlement continued to expand into it, the southern half was devasted and the northern subject to over exploitation.

Another enquiry, again in defence of rights of the crown, was held in 1619 at Alresford. The theme was royal land held by various local landowners at inadequate rents.(52) Sir Richard Norton of Rotherfield (East Tisted) held 263 acres, with cottages, near Oakhanger Pond and Eveley for 13s-6d per annum. It was argued the true rent was £20. Thomas Pescod held 47 acres of Oakwood Marsh (near Blackmore) for 3s-6d, whereas the fair rent was judged to be £7-10s. Similarly John Newlyn, Edward Chase, John Holloway, all of yeoman class, held extensive acreages around Blackmore at derisory rents. The same applied to much of the forest. Forest lands where contributing to the social ascent of various local families and, with loss of grazing rights, the increasing penury of others.

The administration of the forest with rangers, woodwards, regarders and agisters was still in being in 1627(53) but the court proceedings were concerned mainly with the continued spread of cottages, and with unlawful felling. In the attemps to renew the application of forest law, and the assertion of crown rights, local people were required to appeal for their pasture rights in the forest. The claimants extended from the Bishop of Winchester for his manor of Farnham and Wrecclesham in the north to John Biden of Liss Sturmey in the south. Other claimants included representatives of local families already named: Pescod for Oakhanger, Freland for Greatham and Newlyn also for Oakhanger. Henry Hook claimant for Bramshott was verderer, and Robert Tirwitt warden.(54) Tirwitt was still 'lieutenant of the forest' in 1635, the last eyre to be held.

The forest was still divided into five walks each with a forester. Wolmer Forest south was heath except around Linchborough Lodge, but some mixed oak woodland survived in Alice Holt, the Strait being described as 'a wood a mile length, three furlongs broad with building timber and ships timber'. The burning and turf cutting in Wolmer Forest south had destroyed the soil as well as the woodland, but Alice Holt, on heavy clay, survived. The unlawful fellings continued into the Commonwealth as shown by yet another enquiry in 1653.(55) There is no reference throughout the 16th and 17th century to royal hunting, and the forest courts, when held, were concerned with other matters, but there was certainly extensive poaching of surviving game. Poachers over a long period had painted their faces black to avoid being easily seen at night, and there was legislation from 1485 against poachers with 'painted faces'. The most notorious group, at work in the early 18th century was known as the 'Waltham Blacks' (coming from Bishop's Waltham). In 1718 the penalty was £ 50 fine or three years imprisonment. in 1719 the penalty was greatly increased to transportation to America. In 1722, as previously mentioned, the death penalty was re-introduced (from its abrogation in 1217) for poaching with blackened faces, and without benefit of clergy. Four offenders were hung in chains at Reading Assizes.

Modern Wolmer Forest

The 6th report of the Commissioners appointed to enquire into the state and conditions of the woods, forests and land revenues of the crown, published in 1790, can be considered to mark the beginning of modern Wolmer Forest.

The total acreage of the forest was then 15,473 of which 8,694 were crown property. Surviving woodland was concentrated in Alice Holt but no naval timber had been cut in the 18th century until 1777. The mention of naval timber production marks the change of emphasis in usage. Since Queen Anne's reign there had been very little administration and although the surviving red deer had been removed from Wolmer Forest south, Alice Holt was overstocked with 1500 fallow deer. Too much scrub had been cut removing nurse trees and preventing regeneration. Many trees had been ruined by pollarding. The Commissioners advised exclusion of deer and stock from enclosures to permit regeneration in Alice Holt. Scientific foresty commenced with pine and oak plantation in the early 19th century. In Wolmer Forest south only the Linchborough area was seen as fit for woodland. Most was heath, although some coniferous planting had taken place in 1750. Alice Holt is now administered by the Forestry Commission. Wolmer Forest south was finally enclosed in 1864-6, crown land passing into the hands of the Forestry Commission and then leased to the War Office. Wolmer Forest south is now a mixture of heath and plantation used by the Army as a firing range. A paradoxical effect of this usage, as previously discussed, is the survival of a diverse fauna, and a marked increase in the avifauna. The enclosure, by removing commoning rights from large areas, permitted an explosion of building in Liphook, Headley Down and Whitehill, continuing the expansion function of the former forest. Wolmer Forest south is also seen by the planning authorities as an area for the expansion of industries and services to which there are objections in the 'pretty' villages to the south and west. The forest lands are still regarded as a disposable asset, although geology and forest law have combined to preserve here some of the oldest fragments of vegetation with indigenous fauna in south England.

Manuscript references

All Public Record Office, London, except where otherwise specified. Asterisked references are available in transcription. (1) E 32/259 (2) E 101/147/17 (3) E 36/75 (4) E 101/497/10 (5) Pipe Roll for 1190* (6) E 32/158 (7) C 47/11/5 m 19 (8) E 32/161 (9) E 32/310 (10) E 32/339 (11) E 32/259 (12) E 32/261 (13) E

32/311 (14) E 32/161 (15) E 32/310 (16) E 32/261, the evidence is in a reference to poaching in the field of Liss called Spireham. The carcase was carried to the home of Richard atte Burgate. Burgate is in West Liss. (17) will of John Hill of Empshott 1588, Hampshire Record Office 'one acre of land in the middle field of Greatham'. (18) E 32/158 (19) ibid. (20) E 101/147/19 (21) E 32/361 (22) E 134/28 Eliz./Easter 18 (23) E 101/147/17 (24) E 32/261 (25) E 32/310 (26) Rotuli Hundredorum* Record Commissioners, 1812-18, ii p. 224. (27) Itinerary of Edward I, P.R.O. Round Room, typed list (28) E 32/279 (29) Records of wardrobe and household.* edited B.F. Byerly and C.R. Byerly, H.M.S.O. 1977. (30) E 101/561/41 (31) E 101/497/11 (32) E 32/168 (33) E 32/310 (34) E 32/311 (35) E 32/310 (36) E 32/311 (37) E 32/168 (38) E 32/257 (39) E 32/157 and E 32/158 (40) E 133/10/1614 (41) E 32/279 (42) E 32/157 (43) E 32/158 (44) DL 39/1 (45) E 36/75 (46) E 32/261 (47) E 32/168 (48) E 32/310 (49) E 101/619/6 (50) SC 6/983/35 (51) E 101/147/19 (52) E 134/28 Eliz/Easter 18 (53) SP 14/203/2 (54) C 99/3 (55) C 99/1 E 146/3/18.

Other references

Capes, W.W. 1901. Scenes of rural life in Hampshire among the manors of Bramshott 1901.
Commissioners appointed to enquire into the state and condition of woods, forests and land revenues of the crown 1790, 6th report.
Cox, J.C. 1905. The royal forests of England.
Doubleday, H.A. and W. Page editors 1903. Victoria County History of Hampshire, vol. 2.
Meirion-Jones, G.I. 1968. A contribution to the historical geography of north-east Hampshire c. 1600-1850. unpublished M. Phil. thesis University of London.
W. Page editor, 1908. Victoria County History of Hampshire, vol. 3.
Stenton, D.M. 1951. English society in the early Middle Ages.
Turner, G.J. 1899. Select pleas of the forest, Selden Society 13.
White, Gilbert 1789. Natural history and antiquities of Selborne. numerous subsequent editions.

Address of the Author:
Prof. Dr. E.M. Yates
University of London Kings College
Strand London WCZR ZLS

COMPARITIVE STUDIES ON SPECIES DIVERSITY OF PLANT ASSOCIATIONS IN THE U.S.A. AND NORTHERN GERMANY

O. Fränzle

> Many examples of the influence of random events upon natural systems....suggest that instability, in the sense of large fluctuations, may introduce a resilience and a capacity to persist.
>
> *Holling* (1976)

1. Introduction

In ecology the discussion of stability has traditionally tended to equate stability with systems behaviour. Questions of persistence and the probability of extinction are of particular importance in this connection, and hence these measures are also frequently defined as aspects of stability. It should be kept in mind, however, that stability is the ability of a system to return to an equilibrium state after a temporary disturbance, while persistence is a measure of resilience in the sense defined by Holling (1976), i.e. the capacity to absorb changes of state and driving variables and parameters. The more rapidly a system returns to equilibrium, and with the least fluctuation, the more stable it is, and this means, thermodynamically speaking, that stability is coupled with a (relative) minimum of entropy production (Prigogine, 1976).

The analysis of the intricate 'interplay between resilience and stability' (Holling, l.c.) might therefore help to clarify the conflicting views of relationships between diversity and stability of ecological communities. For instance, MacArthur (1955) has argued cogently that stability is proportional to the number of links between species in a trophic web, while Müller (1977) emphasized that less diversified communities might exhibit a higher stability against external disturbances than others richer in species. May's (1973) comparative mathematical analysis of models of interacting populations shows, however, that a relation between increased diversity and stability is not mathematically cogent. Randomly assembled complex systems are in general less stable than less complex ones, and they might fluctuate distinctly more. Since ecosystems are likely to have evolved to a very small subset of all possible sets, however, Mac Arthur's and analogous conclusions might still apply in the real world.

It ensues from a part of these analyses and studies on species diversity of tropical rain forest ecosystems in particular (Fränzle, 1976, 1977) that a system is the more likely to have both low fluctuations and low resilience the more homogeneous its environment in space and time is. This positive relationship between diversity and stability is appropriately illustrated by the Hylaea associations of

113

South America. Here the entropy of the impoverished mature soils of the ferralsol, acrisol, and podzol groups is high in terms of low content and equal distribution of nutrients; consequently the net entropy rates of the vegetation cover must be proportionally low in order to maintain stability of the total ecosystem in the most efficient way. In other words: species diversity of phytocenoses and nutrient content of the sites are negatively correlated provided energy flux rates are high enough to ensure stability by means of efficient entropy flux processes (Fränzle, 1977). It is the purpose of the present article to more precisely determine the realm of validity of this statement which is theoretically conclusive but has been empirically tested so far only with respect to (presumably) climacic tropical lowland associations. It is dedicated to Professor Schmithüsen in grateful acknowledgement of many a fruitful and inspiring discussion in the field of plant geography.

2. Diversity Analyses of Plant Associations

The present article is based on the evaluation of 132 U.S. American and 194 German plant associations defined by modern phytosociological methods which permit a quantitative interpretation with regard to species diversity and related characteristics.

The notation of structural diversity applied is Shannon's entropy

$$H = - \sum_{i=1}^{n} p_i . ld\, p_i \tag{1}$$

with $p_i = p_1, p_2, \dots p_n$ denoting the proportional abundances of the n species in a sample. Information-theoretic notations of the above type are now well established in ecology and systematics (Hill, 1973), and the Shannon measure is given preference to the generalized entropy of order 2 (Renyi, 1961) because it is illuminating in the biophysical context. In addition to H also maximum diversity is determined, defining

$$H_{max} = ld\, n \tag{2}$$

with n = number of species in a sample.

Since these statistics, and in particular the last one, are strongly affected by the presence of rarities a normed abundance measure $\hat{A}$ is further introduced to better characterize the structure of an association:

$$\hat{A} = 100\, n \cdot N^{-1} \tag{3}$$

where n has the above meaning and N is the total number of plants encountered in the sample.

114

The composition of the 132 plant associations analyzed in relation to climate and soils was derived from the following publications: Ahlgren (1960), Billings & Thompson (1957), Bourdeau & Oosting (1959), Bray (1959, 1960), Brown (1958), Buell & Bormann (1955), Buell & Cantlon (1950, 1951), Caplenor (1968), Cottam (1949), Culberson (1958), Dix (1957, 1960), Dix & Butler (1960), Douglas & Ballard (1971), Hurd (1961), Isaak et al. (1959), Kucera (1956), Kucera & Martin (1957), Küchler (1951), Livingston (1952), Maycock (1961), McIntosh (1959), McNaugton (1968), Merkle (1951, 1954, 1962), Mueggler & Harris (1970), Neiland (1958), Potter & Green (1964), Potter & Moir (1961), Ray (1959), Thilenius (1968), West & Ibrahim (1968), White (1967), Whittaker & Niering (1965). The areas sampled represent fairly well the major part of the 35 natural subdivisions of the continental U.S.A. without Alaska as defined by Hammond (1965) and permitted, by means of additional information drawn from other climatological and geomorphological sources, the calculation of the above mentioned diversity parameters and 16 related values characterizing the site structure.

A concise and accurate picture of these site characteristics and their relationship to diversity is provided by the following symmetrical 11×11 matrix showing the statistical quality of the numerical correlations established in terms of significance levels (fig. 1).

Correlations of the variables H, H_{max}, Hu, P, and PD are of particular importance in the present context, hence they are formulated as linear regressions. This notation has the advantage of conciseness and it contributes essentially to reducing the influence of subjective differences in the primary phyto-sociological sampling procedure. These can be assessed, as in the present case, by means of appropriate methods, e.g. redundancy analysis (Lausi, 1972) on the basis of the H, H_{max} and $\hat{A}$ relationships.

Table 1. Correlations of diversity measures with selected site factors.

Correlated variables	Regressions	r^2 (%)
$H \oplus H_{max}$	$H = 0.43\ H_{max} + 2.99$	39.2
$H \oplus PD$	$H \Rightarrow 9.31\ PD + 61.86$	6.6
$\hat{A} \oplus P$	$\hat{A} \Rightarrow 2.62\ P + 60.63$	16.3
$\hat{A} \oplus Hu$	$\hat{A} \Rightarrow 3.52\ Hu - 71.58$	11.9

It ensues from these data that precipitation, measured as water depth, is statistically irrelevant for diversity but exerts a strong control on normed proportional abundance, while humidity is only of fairly limited importance for diversity. A multiple linear regression between diversity, water depth and evaporation shows that the low significance level of the latter relationship is due to the fact that the influence of precipitation on humidity is overcompensated by that of potential

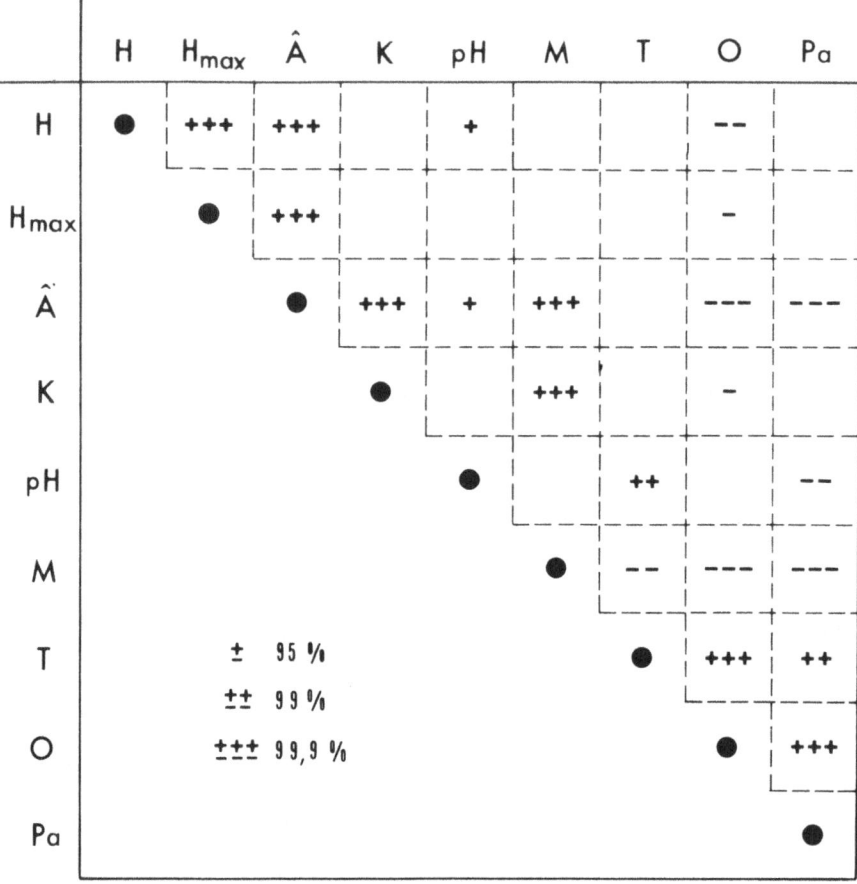

Fig. 1., Symmetrical 11 x 11 matrix of the statistical significance of correlations between diversity measures and selected site qualities (U.S.A.). H = species diversity (Shannon's entropy), H_{max} = maximum diversity, $\hat{A}$ = normed proportional abundance, λ = longitude ω, φ = latitude N, E = class a pan evaporation in cm/a, Hu = humidity (i.e. pricipitation minus evaporation E in cm/a, R = solar radiation in kLy/a, NR = net radiation in kLy/a, PD = number of days with rainfall, P = precipitation in cm/a.

evaporation derived from class A pan data. Hence it may be concluded that the amount of water available for evapotranspiration is of decisive importance for species diversity. This conclusion is further corroborated by the highly significant correlation between diversity and distribution of rainfall in time (variable PD). It merits special attention that neither H nor H_{max} co-variate in a statistically significant manner with $\hat{A}$, and it is important that H_{max} controls only 39.2% of the observed variability of H. This surprisingly low value is indicative of a very moderate average evenness in the sense defined by Pielou (1969) but might, in addition, also hint at a certain degree of subjectivity in sampling.

2.2 The Structure of Plant associations in Northern Germany

The information drawn upon in this section comes from the following publications which describe a varying number of natural or relatively natural associations with their corresponding soils: Buchwald (1950), Bieschke (1969), Füllekrug (1968), Hofmann (1965), Lohmeyer (1951), Müller-Stoll & Krausch (1968), Passarge & Hofmann (1968), Schlottmann (1966), Schlüter (1955), Trautmann & Lohmeyer (1960). Taking into account the different phyto-sociological approaches of these authors the associations are grouped into a 108 samples' subset I (Hofmann, l.c., Passarge & Hofmann, l.c.) and a subset II composed of the remaining 86 associations. With one exception (Füllekrug, l.c.) the 194 associations evaluated grow on soils developed from glacial, glacifluvial and periglacial deposits of the Weichselian and Saalian.

A more comprehensive and yet concise picture of the site characteristics is provided by the following matrices (figs. 2 and 3). Figure 2 summarizes the statistically significant correlations derived from subset I, figure 3 gives more detailed information on the associations of subset II sampled according to the Braun-Blanquet method.

Table 2 summarizes the important correlations in numerical form.

Table 2. Correlations of diversity measures with selected site factors.

Correlated variables			Regressions			$n^2\%$
H_{max}	⊕ H	(3)[1]	H_{max}	= 0.97 H	+ 0.89	96.3
H_{max}	⊕ H	(2)[1]	H_{max}	= 0.97 H	+ 0.56	92.8
H_{max}	⊕ $\hat{A}$	(3)	H_{max}	= 0.74 $\hat{A}$	+ 2.97	36.1
H_{max}	⊕ N	(2)	H_{max}	= 0.11 N	+ 5.11	22.8
H	⊕ $\hat{A}$	(3)	H	= 0.09 $\hat{A}$	+ 2.30	38.4
H	⊕ N	(2)	H	= 0.12 N	+ 4.66	25.8
H	⊕ pH	(3)	H	= 0.14 pH	+ 4.18	4.0
$\hat{A}$	⊕ K	(3)	$\hat{A}$	⇒ 0.74 K	+ 13.70	12.3
$\hat{A}$	⊕ M	(3)	$\hat{A}$	= 0.86 M	+ 21.04	12.3
$\hat{A}$	⊕ H_{max}	(2)	$\hat{A}$	= 0.15 H_{max}	+ 5.29	8.1
$\hat{A}$	⊕ pH	(3)	$\hat{A}$	= 0.97 pH	+ 22.87	4.0

[1] (2) and (3) refer to matrices 2 or 3, respectively.

In comparison to table 1 particular attention must be drawn to the fact that diversity tends to increase with nutrient supply, while soil humidity does not exert a significant influence in this respect in northern Germany.

Cluster analysis (program YHAK by Forst, Köln & Vogel, Kiel) in the form of centroid sorting which yielded distinctly better results than other hierarchical classification procedures tested to this end (single and complete linkage, Ward's algorithm) permitted to subdivide the subsets I and II into 16 or 15 association classes, respectively, on the basis of their diversity and abundance characteristics. Furthermore the information on subset I enabled the generation of 15 site classes in terms of nutrient-moisture combinations. Both classefication results are valid

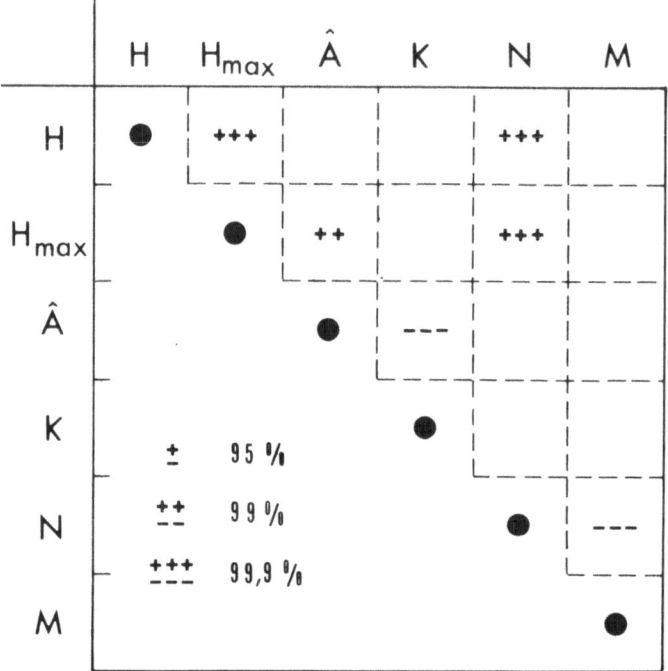

Fig. 2. Symmetrical 6 x 6 matrix of the statistical significance of correlations between diversity measures and selected site qualities (northern Germany, subset I). H, H_{max}, A as in figure 1, K = thermal continentality (Gorczinski formula, 1920), N = nutrient status (8 = rich mineral soil, ..., 1 = poor organic soil), M = soil moisture (1 = dry mineral soil, ..., 8 = swampy organic soil).

on the 5% level of cumulative entropy increase, and can be appropriately combined in the form of a frequency matrix. It gives a concise picture of the association structure in terms of diversity, normed proportional abundance, nutrient status and soil moisture.

This matrix shows that most frequently high diversity coupled with medium to low proportional abundances (classes 12-16) occurs on sites with high nutrient supply and medium to low soil moisture (classes 1-5), whereas the combination of both high diversity and high abundance (classes 5-7) with high nutrient and medium to low moisture status (classes 1-5) is comparatively rare. The following section will show that this result contributes considerably to the interpretation of the regressions tabulated in table 2 and also helps to better understand the relationships listed in table 1.

3. Conclusions

In the light of the above results the validity of the introductory statement that species diversity of phytocenoses and nutrient supply of the sites are negatively

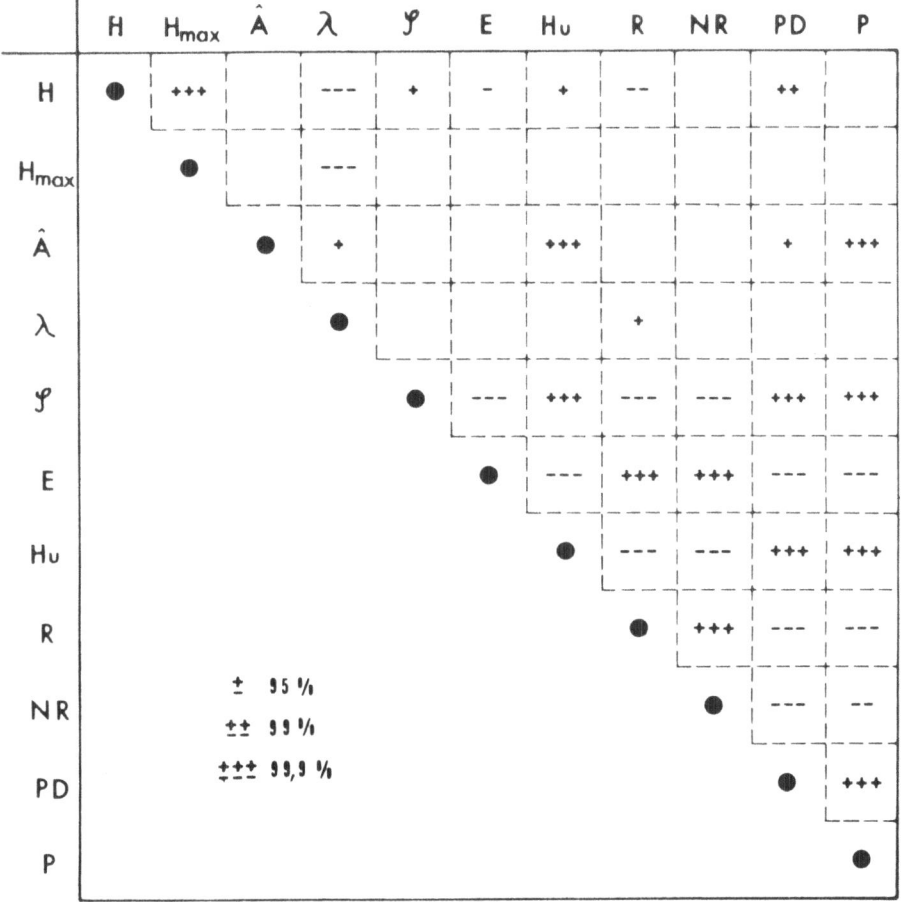

Fig. 3. Symmetrical 9 x 9 matrix of the statistical significance of correlations between diversity measures and selected site qualities (Northern Germany, subset II). H, H_{max}, Â, K, M, as in figure 2, pH = pH value of soil, T = textural classes (1 = sand, ..., 8 = clay), O = quality of organic matter (1 = mild humus, ..., 3 = raw humus, 5 = peaty soil, 6 = peat), Pa = parent material (1 = basic, ..., 5 = acid rock, 6 = peat).

correlated provided energy fluxes are high enough has to be limited to tropical lowlands. Furthermore the analysis of the American associations shows that diversity tends to increase with the amount of water available for evapotranspiration while the German associations clearly indicate a positive influence of nutrient supply on species diversity.

These inverse diversity-nutrient relationships in tropical and ektropical environments on the one hand and the fact that soil moisture controls diversity in the U.S.A. but not in Northern Germany on the other lead to the question if a unifying interpretation of these apparently incoherent results is possible. In this connection it should be recalled that an increase of net primary production or

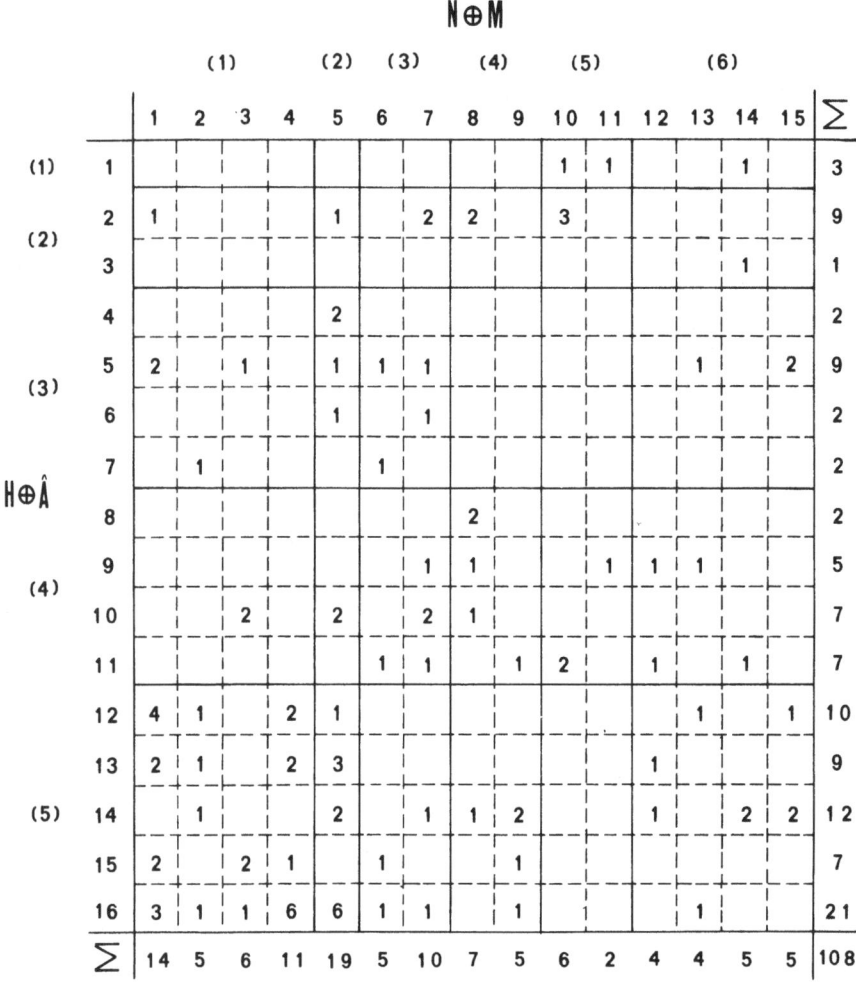

Fig. 4. 15 x 16 frequency matrix of diversity⊕abundance versus nutrient⊕moisture combinations of subset I associations.

biomass with growing soil moisture and elevated CO_2 level of the microclimate does not necessarily imply a corresponding increase in structural diversity (cf. mono- or oligospecific associations). Hence the present article does not refer to well-known physiological plant-water relationships but wishes to stress some more general biophysical aspects which seem to have played a lesser role in the relevant discussion hitherto.

3.1 Some general notions of thermodynamics and dissipative structures

Biophysical aspects of plant-life are basically related to specifications of the second law of thermodynamics as applied to open systems (Prigogine, 1967). Because

120

these systems are in exchange of energy and matter with the outside world their entropy production dS comprises the terms d_iS and d_eS where d_iS denotes the entropy production within the system while d_eS is a flux term describing entropy 'export' into the environment:

$$dS = d_iS + d_eS \tag{4}$$

Only $d_iS \geqslant 0$ but d_eS can also be negative. Identifying entropy with disorder (Boltzmann, 1872), it ensues from equation (4) that an isolated system can only evolve towards greater disorder. For an open system, however, the 'competition' between d_eS and d_iS permits the system, subject to certain boundary conditions, to adopt new states or structures. These are stationery if

$$dS = 0 \text{ or} \tag{5a}$$
$$d_eS = -d_iS \leqslant 0 \tag{5b}$$

respectively.

d_iS can be expressed in terms of thermodynamic 'forces' X_i and rates of irreversible phenomena J_i (Prigogine, l.c.). X_i may be gradients of temperature or concentration; the corresponding 'rates' are then heat flux and chemical reaction rate. Hence

$$\frac{d_iS}{dt} = \sum_{i=1}^{n} J_iX_i \tag{6}$$

Around equilibrium there is a linear relationship between fluxes and forces

$$J_i = \sum_{j=1}^{m} L_{ij}X_j \tag{7}$$

where L_i are specific coefficients, e.g. coefficients of thermal conductivity or diffusion. Provided the reservoirs of energy and matter in the environment of the open system are sufficiently large to remain essentially unchanged, the system can tend to a nonequilibrium stationary state far beyond the domain of linear thermodynamics. This state may be associated with dissipative structures (Glansdorff & Prigogine, 1971), i.e. structures resulting from a dissipation of energy rather than from conservative molecular forces.

3.2 Phytocenoses as dissipative structures

Considering physocenoses from the point of view of stationary dissipative structures the relationship of d_iS and d_eS as expressed in equation (5) and the specific boundary conditions controlling entropy production and flux rates appear to be particularly important. It is a consequence of equation (5b) that, thermodynamically speaking, stability or the capacity to maintain a nonequilibrium steady state is coupled with a (relative) minimum of total entropy production dS. Clearly this can be accomplished by either minimizing d_iS or maximizing d_eS, or a combination of both strategies.

Concentration processes involved in the normal metabolic activities of living systems play an important role in this connection as can be seen from the following equation

$$\Delta G^\circ = R \cdot T \cdot \ln \frac{C_2}{C_1} \tag{8}$$

where ΔG° = difference in standard free energy, $R = 8.31$ J·mol^{-1} K^{-1}, T = temperature in Kelvin and C_2, C_1 = higher or lower thermodynamic concentration, respectively. Changes in concentration are a physical prerequisite for the production of a great many compounds, and an absolutely cogent one if substances are produced whose free energy is higher than that of the corresponding 'raw materials'.

Taking into account that even in the simplest cells the normal metabolic pathways imply several thousand complex chemical reactions which must be coordinated by means of an extremely sophisticated functional network means that biological order in both functional and spatio-temporal respect constitutes a further and most powerful negentropic factor. It characterizes every living system from the sub-microscopic level up to gigantic rainforest biomes like the Amazonian Hylaea.

The effectiveness of these negentropic processes is further enhanced by most efficient entropy fluxes related to the transpiration and nocturnal respiration of plants. The molal entropy of H_2O increases from 63 J.mol^{-1}.K^{-1} (liquid) to 189 J.mol^{-1}.K^{-1} (gas) in the course of evaporation, and CO_2 has a molal entropy of no less than 214 J.mol^{-1}.K^{-1}. Consequently also the reverse process, the photosynthetic fixation of CO_2, is of comparable importance for the negentropy balance.

In the light of these mechanisms the results of the present comparative diversity analyses and those dealing with tropical rainforest associations (Fränzle, 1976) may be given a unified interpretation:
(i) Species diversity is not a monotonous function of the nutrient status of soils. If soils form on basic or intermediate parent materials pedogenic nutrient supply and species diversity may both increase during periods of $10^3 - 10^4$ (10^5) years,

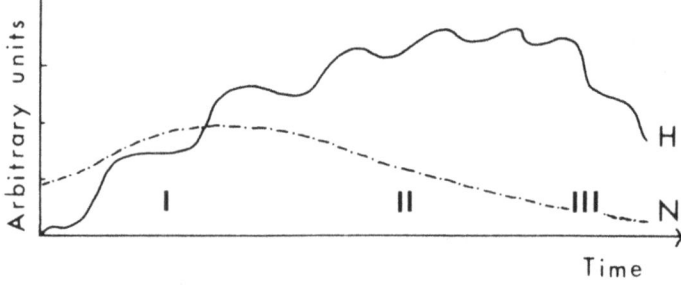

Fig. 5. Schematic representation of evolutionary trends of species diversity and pedogenic nutrient supply with non-linear time scale. I, II, III indicate the phases described in text.

provided the water and energy factors are not limiting. In figure 5 the period of ascending evolution is termed phase I. Shorter fluctuations of variable but generally decreasing amplitude are likely to be superimposed on this long-term trend.

In phase II which can last for several hundred thousand years soils degrade in regard to nutrient supply but species diversity keeps increasing for negentropic reasons (cf. the tropical rainforest stands in comparison to the ektropical associations described here). Clearly such an evolution towards and including the following phase requires a sufficient degree of climatic and geomorphic stability or else the continuity of pedogenesis will be interrupted by truncation processes that rejuvenate the soils. Phase III marks the eventual and comparatively acute decrease in diversity of phytocenoses once the nutrient status of soils has fallen below a critical level.

(ii) Spatial heterogeneity as a characteristic feature of the sites during phase I results in variability in numbers of species and plant individuals. With this variability the associations can simultaneously retain genetic and behavioural types that can maintain their existence in low populations together with others which can capitalize on opportunities for dramatic increase.

The more homogeneous the environment in space and time (i.e. during phase II, and probably also phase III), the more likely is the system to have low fluctuations i.e. high stability. Tropical rainforests represent climatically buffered and (largely) self-contained systems with relatively low (natural) variability on highly impoverished soils of the ferralsol, acrisol and podzol classes. Temporary external disturbances are consequently likely to affect a proportionally higher number of species than in phase I populations, where the interspecific differences in ecological potency are usually distinctly higher. Hence resilience of mature biocenoses is low during phase II. On the contrary biocenoses may be highly resilient in phase III of evolution, although unstable.

4. Acknowledgements

I am grateful to Miss M. Bruns and Mr. W. Matzick for help in compilation and primary evaluation of phyto-sociological data in respect to diversity. I am indebted to Miss A. Ingenpass and Mr. E.-D. Wallrodt for support in cluster analysis.

Zusammenfassung

Ausgehend von der Tatsache, daß eine klimatische Climaxgesellschaft mit ihren Böden ein stationäres Makrosystem hohen Ordnungsgrades bildet, läßt sich für die tropischen Tieflandsregenwälder zeigen, daß die geringen Nährstoffreserven ihrer Böden in Verbindung mit den hohen Energieflüssen die außerordentlich hohe Arten-Diversität bedingen. Vergleichende statistische Untersuchungen der als Informationsentropie (Shannon) definierten Diversität von mehr als dreihundert pflanzensoziologisch aufgenommenen Beständen in den USA und Norddeutschland ergänzen diese Befunde und präzisieren ihren Gültigkeitsrahmen.

Aus dem Vergleich der Bestandesstruktur in diesen florengeschichtlich und pedogenetisch gut bekannten Untersuchungsräumen ergibt sich, daß die Diversität natürlicher (bzw. naturnaher) Bestände in den USA mit der Menge pflanzenverfügbaren Wassers, in Norddeutschland mit dem Nährstoffgehalt der jungquartären Böden wächst. Die Gesamtheit dieser Befunde aus den Tropen und Ektropen lassen sich unter Berücksichtigung der molaren Entropie des Wassers und Kohlendioxids einer vereinheitlichten Deutung zuführen und in einem Entwicklungsmodell zusammenfassen, welches drei Phasen der Boden- und Vegetationsentwicklung unterscheidet. In der ersten, die 10^3-10^4 (10^5) Jahre dauert, steigt die verwitterungsbedingte Nährstoffversorgung der Böden auf basischen bis intermediären Ausgangsgesteinen bis zu einem Maximum an; wie die Struktur mitteleuropäischer und nordamerikanischer Bestände zeigt, nimmt die Artendiversität gleichzeitig unter Schwankungen zu. Bei relativer Klimakonstanz und geomorphodynamischer Stabilität schließt sich die Phase II an, die mehrere hunderttausend Jahre währen kann. Sie ist durch weitergehende Diversitätszunahme bei gleichzeitiger Nährstoffverarmung der Böden gekennzeichnet und wird durch entwicklungsgeschichtlich alte tropische Regenwälder bzw. immergrüne Subtropenwälder im Bereich klimastabiler passatischer Steigungsregen repräsentiert. Phase III kennzeichnet die verhältnismäßig rasch erfolgende strukturelle Verarmung der Bestände, sobald der Nährstoffpegel unter einen systemspezifischen Grenzwert gesunken ist. Die negentropischen Prozesse, die für Phase II charakteristisch sind, vermögen nun das System nur noch auf einem durch relativ niedrige Diversität gekennzeichneten Niveau zu stabilisieren. Abschließend werden die phasenspezifischen Unterschiede der Stabilität und Belastbarkeit tropischer und ektropischer Pflanzenbestände beschrieben.

5. References

Ahlgren. C.E. 1960. Some effects of fire on reproduction and growth of vegetation in Northeastern Minnesota. *Ecology* 41: 431-445.

Billings, W.D. & Thompson, J.H. 1957. Composition of a stand of old bristlecone pines in den White Mountains of California. *Ecology* 38: 158-160.

Bourdeau, P.F. & Oosting, H.J. 1959. Maritime live oak forest in North Carolina. *Ecology* 40: 148-152.

Braun-Blanquet, J. 1964. Pflanzensoziologie. Grundzüge der Vegetationskunde. Wien.

Bray, J.R. 1956. A study of mutual occurrence of plant species. *Ecology* 37: 21-28·

Bray, J.R. 1960. The composition of savanna vegetation in Wisconsin. *Ecology* 41: 721-732.

Brown, H.E. 1958. Gambel oak in west-central Colorado. *Ecology* 39: 317-327.

Bruns, M. 1974. Vergleichende vegetationsgeographische Untersuchungen von Pflanzenformationen in Gebieten gleicher Strahlungsbilanz (unpubl.)

Buchwald, W. 1951. Wald- und Forstgesellchaften der Revierförsterei Diensthoop, Forstamt Syke bei Bremen, in: Angewandte Pflanzensoziologie. 1.

Buell, M.F. & Cantlon, J.E. 1950. A study of two communities of the New Yersey Pine Barrens and a comparison of methods. *Ecology* 31: 567-586.

Buell, M.F. & Cantlon, J.E. 1951. A study of two forest stands in Minnesota with an interpretation of the prairie-forest margin. *Ecology* 32: 294-316.

Buell, M.F. & Bormann, F.H. 1955. Deciduous forests of Ponemah Point, Red Lake Indian Reservation, Minnesota. *Ecology* 36: 646-658.

Caplenor, D. 1968. Forest composition on loessal and non-loessal soils in west-central Mississippi.

Caplenor, D. 1968. Forest composition on loessal and non-loessal soils in west-central Mississippi. *Ecology* 49: 322-331.

Cottam, G. 1949. The phytosociology of an oak wood in south-western Wisconsin. *Ecology* 30: 271-287.

Culberson, W.L. 1958. Variation in the pine-inhabiting vegetation of North Carolina. *Ecology* 39: 23-28.

Dierschke, H. 1969. Natürliche und naturnahe Vegetation in den Tälern der Böhme und Fintau in der Lüneburger Heide. *Mitt. der flor.-soziol. AG. Nf.* 14: 377-397.

Dix, R.L. 1957. Sugar maple in climax forests at Washington D.C. *Ecology* 38: 663-665.

Dix, R.L. 1960. The effects of burning on the mulch structure and species composition of grasslands in Western North Dakota. *Ecology* 41: 49-56.

Dix, R.L. & Butler, J.E. 1960. A phytosociological study of a small prairie in Wisconsin. *Ecology* 41: 316-327.

Douglas, G.W. & Ballard, T.M. 1971. Effects of fire on alpine plant communities in the North Cascades, Washington. *Ecology* 52: 1058-1064.

Fränzle, O. 1976. Der Wasserhaushalt des amazonischen Regenwaldes und seine Beeinflussung durch den Menschen. *Amazonia* 6: 21-46.

Fränzle, O. 1977. Biophysical aspects of species diversity in tropical rain forest ecosystems. Biogeographica 8: 69-83.

Füllekrug, E. 1968. Die Waldgesellschaften an der Schanze bei Bad Gandersheim und ihre räumliche Gliederung. *Vegetatio* 15: 51-76.

Glansdorff, P. & Prigogine, J. 1971. Thermodynamic Theory of Structure, Stability and Fluctuations. New York.

Gorczynski, W. 1920. Sur le calcul du degré de continentalisme et son application dans la climatologie. *Geogr. Annaler*, 2: 324-331.

Hill, M.O. 1973. Diversity and evenness: A unifying notation and its consequences. *Ecology* 54: 427-432.

Hofmann, G. 1965. Waldgesellschaften der östlichen Uckermark. Feddes Repertorium, Beiheft 142: 133-202.

Holling, C.S. 1976. Resilience and stability of ecosystems. In: Jantsch, E. & Waddington, C.H. (Eds.): Evolution and Consciousness: 73-92.

Hurd, R.M. 1961. Grassland vegetation in the Big Horn Mountains, Wyoming. *Ecology* 42: 459-467.

Isaak, D., Marhall, W.H. & Buell, M.F. 1959. A record of reverse plant succession in a tamarack bog. *Ecology* 40: 317-320.

Kucera, C.L. 1956. Grazing effects on composition of virgin prairie in North-Central Missouri. *Ecology* 37: 389-391.

Kucera, C.L. & Martin, S.C. 1957. Vegetation and soil relationships in the Glade Region of the Southwestern Missouri Ozarks. *Ecology* 38: 285-291.

Küchler, A.W. 1951. The relation between classifying and mapping vegetation. *Ecology* 32: 275-283.

Livingston, R.B. 1952. Relict true prairie communities in Central Colorado. *Ecology* 33: 72-86.

Lohmeyer, W. 1951. Die Pflanzengesellschaften der Eilenriede bei Hannover. Angewandte Pflanzensoziologie.

Lotspeich, F.B., Secor, J.B., Okazaki, R. & Smith, H.W. 1961. Vegetation as a soil-forming factor on the Quillayute physiographic unit in Western Clallam County. *Ecology* 42: 58-68.

Mac Arthur, R. 1955. Fluctuations of animal populations and a measure of community stability. *Ecology* 36: 533-536.

Matzick, W. 1975. Untersuchungen über die floristische Diversität nordamerikanischer Bestände in Abhängigkeit von Klima und Boden. (unpubl.)

May, R.M. 1973. Model Ecosystems. Princeton, N.J.

Maycock, P.F. 1961. The spruce-fir forest of the Keweenaw Peninsula, Northern Michigan. *Ecology* 42: 357-365.

McIntosh, R.P. 1959. Presence and cover in pitch pine-oak stands of the Shawangunk Mountains, New York. *Ecology* 40: 482-485.

McNaughton, S.J. 1968. Structure and function in California grasslands. *Ecology* 49: 962-972.

Merkle, J. 1951. An analysis of the plant communities of Mary's Peak, Western Oregon. *Ecology* 32: 618-640.

Merkle, J. 1954. An analysis of the spruce-fir community on the Kaibab Plateau, Arizona. *Ecology* 35: 316-322.

Merkle, J. 1962. Plant communities of Grand Canyon area, Arizona. *Ecology* 43: 698-711.

Mueggler, W.F. & Harris, C.A. 1970. Some vegetation and soil characteristics of mountain grasslands in Central Idaho. *Ecology* 50: 671-678.

Müller, P. 1977. Die Belastbarkeit von Ökosystemen. Mitt. 8, Schwerpunkt für Biogeographie der Universität des Saarlandes. Saarbrücken.

Müller-Stoll, W.R. & Krausch, H.-D. 1968. Der azidophile Kiefern-Traubeneichenwald und seine Kontaktgesellschaften in Mittel-Brandenburg, in: *Mitt. d. Flor.-soziol. Arbeitsgemeinschaft N.F.* 13: 101-121.

Neiland, B.J. 1958. Forest and adjacent burn in the Tillamook burn area of Northwestern Oregon. *Ecology* 39: 660-671.

Passarge, H. & Hofmannn, G. 1968. Pflanzengesellschaften des nord-ostdeutschen Flachlandes II. Pflanzensoziologie 16.

Pielou, E.C. 1969. An Introduction to Mathematical Ecology. New York.

Potter, L.D. & Green, D.L. 1964. Ecology of **ponderosa** pine in western North Dakota. *Ecology* 45: 10-23.

Potter, L.D. & Ross Moir, D. 1961. Phytosociological study of burned decidous woods. Turtle Mountains North Dakota. *Ecology* 42: 468-480.

Prigogine, J. 1967. Thermodynamics of Irreversible Processes, 3rd ed. New York.

Prigogine, J. 1976. Order through fluctuation: Self-organization and social system: In: Jantsch, E. & Waddington, C.H. (Eds.): Evolution and Consciousness: 93-133.

Ray, R. 1959. A phytosociological analysis of the tallgrass prairie in Northeastern Oklahoma. *Ecology* 40: 255-261.

Rényi, A. 1961. On measures of entropy and information. In: J. Neyman (Ed.) 4th Berkeley Symposium on Mathematical Statistics and Probability: 547-561.

Schlottmann, C.B. 1966. Die Pflanzengesellschaften des Gaarder Bauernwaldes (Kreis Südtondern), Mitt. d. AG f. Floristik in S.-H. und Hamburg 14, Kiel.

Schlüter, H. 1955. Das Naturschutzgebiet Strausberg. Feddes Repertorium, Beih. 135: 260-350.

Thilenius, J.F. 1968. The *Quercus garryana* forests of the Willamette Valley, Oregon. *Ecology* 49: 1124-1133.

Trautmann, W. & Lohmeyer, W. 1960. Gehölzgesellschaften in der Flußaue der mittleren Ems, *Mitt. d. floristisch-soziologischen Arbeitsgemeinschaft N.F.* 8: 227-262.

West, N.E. & Ibrahim, K.I. 1968. Soil-vegetation relationship in the shadscale zone of southeastern Utah. *Ecology* 49: 445-456.

White, K.L. 1967. Native bunchgrass (*Stipa pulchra*) on Hastings Reservation, California. *Ecology* 48: 949-955.

Whittaker, R.H. & Niering, W.A. 1965. Vegetation of the Santa Catalina Mountains, Arizona. *Ecology* 46: 429-452.

Address of the author: Prof. Dr. Otto Fränzle,
                          Geographisches Institut der Universität,
                          Kiel, Federal Republic of Germany

GEOGRAPHICAL LANDSCAPE AND MAP CONSIDERATIONS TO SOME PROBLEMS BETWEEN GEOECOLOGY AND CARTOGRAPHY

Carl Rathjens

The discussion arisen around the conception of the geographical landscape (Landschaft) — and in close connection with it also around regional geography (Länderkunde) — has been led with animated intensity during the last years. The term landscape is used here in the meaning of the definition given to the German word 'Landschaft' by German geographers: see also Landschaftskunde (J. Schmithüsen 1976), Landschaftslehre, Landschaftsforschung (E. Neef 1967), Landschaftsökologie (H. Leser 1976) a. s. o. The difficulty of translating this word into other languages has been realized and has become a matter of repeated and often antagonistic argumentations. However it seems to be reasonable to make known also foreign readers with some aspects of a central problem in German geography.

In the course of this discussion of landscape conception in geography numerous essential contributions have been acquired and produced. Some have been substantiating and defending the conception of an integrated geographical landscape and of the synthesis of regional geography, because these terms would be apparently indispensable, if the unity of geographical science should be maintained. But some others also have been rejecting the conception of landscape as appearing unacceptable because of theoretical and methodical reasons. At this place it is not necessary to compile a survey of new literature concerning the discussion of one of these most aggravating modern problems of geography. Just recently J. Schmithüsen (1976) has achieved a very essential contribution to the wide range of questions which have been raised here. Therefore it seems not to be necessary for this paper to substantiate in detail why geography of landscape (Landschaftskunde) and regional geography (Länderkunde) are connected inseparably in this scientific discussion. If the term of landscape is abandoned the rejection of regional geography, with the scientific purpose to identify, to delimitate and to describe geographical regions within given sections of the surface of the globe or the geosphere, must be the consequence. Fortunately we are able to state that among the many contributions to this discussion the serious efforts for a progress of knowledge are predominating far against comments which are prepossessed, polemical or ideologically fixed.

For the understanding of the international state of discussion the remark must be allowed here that one of the sharpest discrepances within geography in the nearest past seems to become settled again in these days. Also geographers in the countries of the socialistic Eastern part of the world have returned to the conception of the unity of geography, that means to a complex subject of geographical

127

research in the so-called landscape and to a regional geography; this is demonstrated today in all internationally visited congresses. The pronounced separation between physical and economical geography with its negative consequences for regional geography is considered there today as dogmatical-stalinistic and is condemned with hard words (f. i. J.G. Sauškin 1966).

In view of this stage of discussions it is astonishing to see that up to now nearly all arguments seem to be lacking which might be contributed by cartography as a science and might come out of the study of topographical, geographical and thematic maps, with the purpose of intensifying the problems here in question.

The reason for this omission is lying apparently in the following fact: very many geographers are working with maps, are using topographical maps for orientation and for fixing observations in the terrain and of surveys of other kind and are putting down their results in thematic maps and cartograms; but only very few geographers are dealing thoroughly with cartography as a scientific discipline. Cartography has developed already since a long time with own efforts into a separate branch of science. The technology of surveying and reproducing maps has become more and more complicated. Generally is has become much less possible for geographers to maintain a review on the more recent development of cartography. Moreover the technical perfection of the reproduction of maps has been placed today into the foreground and has pushed aside the questions of the substantial configuration of maps. But only just in this last point the geographer now as before might and should join in the discussion in a way in which he might exert some influence. Geodesy is predominant today in cartographical matters, the geographical concerns in cartography have receded more into the background.

Nevertheless the fact is too easily forgotten that cartography originally has emerged from geography and that both these disciplines are in possession of the same subject of their scientific efforts. In both sciences the concept of landscape is playing an essential role. In contrast to the discussion in geography nobody in cartography ever has taken offence at the fact that the large-scaled topographical map is a legitimate descendant of the perspective landscape-painting of the early modern times (E. Imhof 1968, see also J. Schmithüsen 1973), — whereas the small-scaled maps are owning a much older root in the route and naval maps.

Cartography normally is defining its subject not only as the surface of the earth. It is speaking often and largely also of the landscape, and even in the same meaning as also geography is using this word. For this fact many evidences can be found. As examples there are to be mentioned the definitions of R. Finsterwalder (1951) for the terms cartography and map, and the 'Allgemeine Kartenkunde' (General Cartography) by W. Bormann (1954) who is referring explicitly to the definitions of R. Finsterwalder. Also in the same sense for instance the Soviet Russian cartographer K.A. Salistchev (1967) is regarding the geographical contents of the earth surface, when he is defining maps as reduced, generalized, mathematically determined representations of the surface of the earth, transferred into plane, showing the distribution, the situation and the connections of the different natural and social phenomena, which have to be selected and characterized corresponding to the respective purpose of each map.

The near relationship of these conceptions to those of geography is becoming especially apparent, if they are compared with the conceptions of the geodetical direction within cartography. So for instance V. Heissler & G. Hake (1970) are calling cartography the science and technique of graphical interpretation of spatially distributed dates. In this definition nothing can be feeled any more of a conception of geographical landscape. Only a short time ago E. Arnberger (1970) has made objections to the supposition geography and cartography having the same aims of investigation. He argued that geography is dealing with landscape, cartography with the graphic representation of landscape. This certainly is of formal correctness, but it becomes a question of secondary importance, if it is calculated that in the first case prominence is given to the object of investigation, in the second case to the operation of investigation.

The preponderance of technical tasks and problems as formulated by the geodetical orientation within cartography, has led to the result that geography is not exerting the necessary influence any more on the ways how cartography is organizing their common subjects and aims of investigation. Geography is not endeavouring enough for a cooperation in those areas where it seems yet to be possible up to now. This is concerning especially the large-scaled maps of the original survey and the topographical maps which are considered today by many geographers only as a useful tool for their own specific work and are only critized from time to time in matters of serviceability. But only very little efforts are made regularly to improve and to develop maps as important base of geographical work. This has not always been the same situation and must not be so necessarily, as may be documented by some names of German geographers like A. Penck, W. Behrmann, M. Eckert-Greifendorff, H. Louis, E. Lehmann, E. Meynen, W. Pillewizer a.o. Especially great is the interest of geomorphology in the large-scaled maps. The German working group for the topographical-geomorphological map examples 1 : 25.000 (Arbeitskreis für die topographisch-geomorphologischen Kartenproben) has demonstrated this. Its results have been published by H. Louis & W. Hofmann (1968 – 1974) in a homogeneous form. – see also my review of this publication (C. Rathjens 1976). In a very good cooperation between geographers and land surveyors over many years the working group has cared for the interest of geomorphology in a correct and clearly understandable representation of the relief of large-scaled maps, and in the same time it has created a cartographical set of good examples of geomorphological landscape types for teaching and demonstrating purposes.

Unfortunately however this working group has found no imitation in this respect in other branches of German geography up to now. It is left therefore to theoretical considerations how a nearer contact to cartography could work in other fields of geography. Cooperation between cartography and the geography of settlement and of agriculture might be successful in the same kind and could help to improve the topographical maps of large scale. The representation of land use in the topographical maps is insufficiently differentiated in most cases. Therefore some considerations have been made already for improvement (C. Rathjens 1957), but have been disregarded up to now. The Atlas of the German agricultural landscape (Atlas der deutschen Agrarlandschaft) edited by E. Otremba (since 1962) did not produce any consequences in the cartographical field. The experimental study of G. Schulz (1969) to design optimally the geographical substance of the topographical map 1 : 25.000 in an example from the Saarland, especially by a much greater differentiation of the forest vegetation and of the areas used by agriculture, apparently has found no wider resonance up to now. In a similar way Chr. Herrmann (1972) has made suggestions to improve the small-scaled topographical map 1 : 500.000 by a combination of hatching and surface colours. Today it might be quite possible already to imagine a useful cooperation between cartography and a geographically orientated ecology of landscape, if and as far as the latter is interested in a well-balanced reproduction of the structures of relevant geofactors in the map (K. Breitfeld 1967). Also from the project of classification of natural regions (Naturräumliche Gliederung), which has been supported decisively by J. Schmithüsen, regrettably no critique of the underlying topographical map material and no discussion with cartography has been developed, though all depends here essentially whether regions equally equipped in natural respect can be differentiated clearly also in their cartographical contents and expression.

The topographical or original cartography has tried from the beginning to comprehend the contents and the characteristics of all those features which are put together today by modern geography under the term of landscape. This cartography has done – within the limits of its possibilities – in a certain section of the earth surface or geosphere as obviously and as completely as possible. Cartography has advanced in this process during a long development from a pictorial reproduction of certain features of the surface in a perspective view, step by step to more abstract and symbolized representations in an exact and true to scale vertical projection. But never the claim has been abandoned that the large-scaled topographical map is and has to be a copy of the spatial or geographical reality. The air photography developed since the beginning of this century with its great correspondences to the map substance has showed that the aim of accuracy in reality has been preserved by cartography in its essential features to this day. Only the representation of the contour lines for the third dimension of the earth surface has to be excepted in this respect; instead of an immediate vivid legibility it has put a mathematical abstraction which is accentuating the horizontal lineation instead of the vertical component of the relief. From this abstract picture the plastical impression and a more exact imagination of the third dimension may be gained only by a learning process of the map reader.

In recent times the satellite photographs of the earth taken from a great altitude have brought evidence that even the small-scaled physical-geographical representations of the atlas cartography – excepting the often quite unnatural colourations of the altitudinal zones – have not deviated very far from the aim of a realistic reproduction of the earth surface. More modern atlasses are endeavoured to substitute the traditional colour gradation of the relief for the true soil colouration conditioned by climate and vegetation. So deserts lying in low land must not appear any more in green colours. G. Schulz (1974) has developed a concept for a renovated atlas of the world in which these criteria of greater nearness to nature have found consideration.

The topographical map is combining the representation of the relief, of the hydrological network, of certain features of the natural vegetation and of the human land use, of the settlements, of the traffic systems and of especial, economically important installations into a typical structural pattern. This pattern is provided with a quite characteristical expression for each region of the earth surface. To recognize such patterns of different geofactors, to understand them and to delimitate one from another, also modern geography is regarding as one of its most important tasks. Landscape geography (Landschaftskunde) in the first line wants to comprehend such patterns in their effective structures and to typify and to classify them, while regional geography (Landes- und Länderkunde) is describing and explaining the patterns as individual spatial figures and is inserting them into the connection of greater geographical regions. Between the patterns of the topographical map and of landscape geography there is an essential difference, though in the patterns of the map the representation necessarily is confined to static and physionomically visible elements of the geosphere, while patterns landscape geography is dealing with may include the total combination of geofactors in its dynamic effectiveness, in cultural landscapes (Kulturlandschaften) also with all influences of the human intellect and will of organization. Nevertheless the fundamentals of the patterns and especially their mutual limitations will remain the same in both cases. Therefore it is allowed to refer to the topographical map as an evidence for the reality of geographical landscapes. A subject not existing in reality would not be reproducable as a characteristic pattern of a topographical map. From this fact there may be derived the right and the duty of landscape geography to cooperate in the production of the large-scaled topographical map, even if this task scarcely has been made use of up to now. With analogous modifications this right and duty should be accepted also for the successive map scales until the products of the atlas cartography are reached.

Also the historical dimension of the landscape development is coming into its own in cartography. Since exact maps are produced the respective conditions of geographical landscapes have been realized and fixed in certain periods, even with minor retardations caused by the renovation of the maps. Therefore the comparison of older and more recent maps is allowing in many cases, especially since the Nineteenth century, to observe the changing of cultural landscapes quite exactly.

Less easily to answer is the question of the relations of geographical landscape and of the various products of the thematic cartography. The question should be touched here only very shortly. For thematic maps their fundament of reference (Beziehungsgrundlage) has priority, allowing to put the cartographical subject into the right relation to the topographical situation. On the base of several suggestions made by H. Louis in this respect H. Kremling (1970) has reflected on this fundament of reference. Only with a sufficient fundament of reference the thematic map may accomplish a contribution to the spatial differentiation of geographical regions. Also in this field there are maps representing patterns of landscape, letting recognize structures of landscape and helping limitate regions of a certain landscape type. We have to think here in the first line for maps of the surface configuration for which a good topographical fundament of reference is unrevocable, for maps of the natural vegetation and of the agricultural land use in the problems of which also J. Schmithüsen has been engaged repeatedly. But to be specified are here also maps of the forms of settlements, of some economical formations and so on. In addition there are to be mentioned the more abstract representations of climatical regions and of singular climatical factors, of density of population and other kinds of thematic maps.

Among the small-scaled representations of atlasses as an exemplary model the atlas of the world 'Die Staaten der Erde und ihre Wirtschaft' created under the direction of E. Lehmann (1964) has to be enumerated here; it is confronting physical and economical maps respectively and is delineating the economical landscapes of the earth in large areas by covering their surfaces totally with colours and conventional signs. Many examples of atlasses for planning purposes are to be consulted for reliable affirmations, even if they are differentiating their contents often in a much too analutical way. The early examples of the German landscape atlasses, in accordance with the model of the atlas of Niedersachsen conceived by K. Brüning (1934), have achieved more in this respect than the regional atlasses published lateron in large numbers. On the other hand conclusions for the spatial differentiation of landscape may be drawn mostly only very indirectly from maps of political territories and structures or from maps of social-geographical circumstances. Very many thematic maps are nearly sterile for landscape geography, especially those representing processes of movement and of changes.

Also landscape geography and geoecology have begun to develop own thematic maps for their own requirements. A very instructive classification of ecological maps has been made by H. Leser (1974, 1976). On one side there are maps relying totally on the contents and the patterns of the topographical map. They are satisfied with limitations, for instance the maps of the spatial differentiation of natural regions the importance of which for geoecology has been underlined by H.-J. Klink (1972). On the other side we find also maps where colours of surface and specific signatures are put into the foreground and where the topographical fundament of reference is put into the background, as in the large-scaled map of the physiotops in the region of the plain of Brilon by L. Finke (1971). One of the best examples of ecological maps of this direction has been produced by D. Werner (1973) for the volcanic region of the Aetna. The decision which kind of cartographic representation might be chosen will depend certainly in most cases on the fact if a landscape unity is to be characterized alone or predominantly by the pattern of the topographical elements of the map in question or if further material is requested for this purpose, for instance out of the sphere of the underlying bed-rock, of the soils, of the vegetation cover, of the differentiated land use and so on.

If we are accepting this narrow interrelation between geographical landscape and map, then there are some essential consequences. If it comes true that maps are representing geographical landscapes as types and as spatial structures under certain assumptions and therefore are substantiating their existence, then they should be considered by geography not only as working media but should be taken into account again more intensely. For this a series of measures is necessary. Already in the training of geographers the occupation with topographical maps should play again a much greater role. In this connection one should not concentrate any more on the problems of drawing and producing maps which are cared for today by the cartographical science and technology in their own responsibility. Much more important is the geographical interpretation of maps which should be pursued in two directions: a critical examination of the possibilities of representing the given geographical contents and an evaluation of the cartographical patterns with regard to their contribution for understanding and differentiating geographical landscapes. It stands to reason that a fundamental knowledge of cartographical technics is necessary for these aims.

For geography as far as it is working scientifically there is resulting the duty to a narrow cooperation with cartography which may be able to supply it, together with air photography, with exact conceptions of the structure and spatial distribution of geographical landscapes. This is so much the more necessary as the mechanization of cartographical technology deriving from geodesy and progressing permanently is threatening to wipe out the expression of topographical maps as characteristic patterns of geographical landscapes. If the spatial distribution of dates within the map is not destined any more by an idea of the structure of a landscape, but only by mechanical methods, then the topographical map will become lost for geography also as a working medium or will at least forfeit much of its value. Good maps strong in its expression and with large contents of information are indispensable for geography of landscape and for regional geography.

Zusammenfassung

Der Beitrag befasst sich mit einigen Beziehungen zwischen der geographischen Landschaft und der Karte, insbesondere der topographischen Karte, die ein Abbild der räumlichen Wirklichkeit darstellt und deren Muster Typen geographischer Landschaften wiedergeben. Karten können daher als Beweismittel für die Existenz geographischer Landschaften herangezogen werden. Umgekehrt ergibt sich daraus für die Geographie das Recht und die Pflicht, stärker an der inhaltlichen Gestaltung der Karten mitzuwirken.

Summary

This contribution is investigating some interrelations between geographical landscape and map, especially the topographical map which gives an image of the spatial reality and with its patterns represents types of geographical landscapes. Therefore maps may be referred to as evidence for the existence of geographical landscapes. Vice versa from this for geography the right and the duty is resulting to cooperate more than today in the substantial configuration of maps.

Literature

Arnberger, E. 1970. Die Kartographie als Wissenschaft und ihre Beziehungen zur Geographie und Geodäsie. Grundsatzfragen der Kartographie, herausgeg. v. d. Österr. *Geogr. Ges.*: 1-28.

Bormann, W. 1954. Allgemeine Kartenkunde. Kartogr. Schriftenreihe, 1, Lahr/Schwarzwald.

Breitfeld, K. 1967. Ein Vorschlag zur einheitlichen Gestaltung ökologisch-geographischer Karten. *Wiss. Zeitschr. Techn. Univ. Dresden*, 16: 891-895.

Brüning, K. 1934. Atlas von Niedersachsen. Oldenburg.

Finke, L. 1971. Die Verwertbarkeit der Bodenschätzungsergebnisse für die Landschaftsökologie, dargestellt am Beispiel der Briloner Hochfläche. Bochumer Geogr. Arbeiten, 10.

Finke, L. 1971. Landschaftsökologie als Angewandte Geographie. *Ber. z. dt. Landesk.*, 45: 167-182.

Finsterwalder, R. 1951. Begriffe Kartographie und Karte. Geogr. Taschenbuch 1951/52: 408-411.

Heissler, V. 1970. Kartograpie I. 4. Aufl., bearb. v. G. Hake. Sammlung Göschen, Berlin.

Herrmann, Chr. 1972. Studie zu einer naturähnlichen topographischen Karte 1 : 500.000 (Kombination von Schräglichtschattierung mit Oberflächenbedeckungsfarben). Diss. Univ. Zürich.

Imhof, E. 1968. Landkartenkunst gestern, heute, morgen. Zürich.

Klink, H.-J. 1972. Geoökologie und naturräumliche Gliederung – Grundlagen der Umweltforschung. *Geogr. Rundschau*, 24: 7-18.

Kremling, H. 1970. Die Beziehungsgrundlage in thematischen Karten in ihrem Verhältnis zum Kartengegenstand. Münchner Geogr. Abhandl., 2.

Lehmann, E. 1964. Weltatlas. Die Staaten der Erde und ihre Wirtschaft. 7. Aufl., Leipzig.

Leser, H. 1974. Thematische und angewandte Karten in Landschaftökologie und Umweltschutz. Verhandl. 39. Dt. Geographentag Kassel, Wiesbaden: 466-480.

Leser, H. 1976. Landschaftsökologie. Stuttgart.

Louis, H. & Hofmann, W. 1968-1975. Landformen im Kartenbild. Topographisch-geomorphologische Kartenproben 1 : 25.000. Braunschweig.

Meynen, E. & Schmithüsen, J. 1953-1962. Handbuch der naturräumlichen Gliederung Deutschlands. Bad Godesberg.

Neef, E. 1967. Die theoretischen Grundlagen der Landschaftslehre. Gotha.

Otremba, E. seit 1962. Atlas der deutschen Agrarlandschaft. Wiesbaden.

Rathjens, C. 1957. Über die Darstellung der Bodennutzung in unseren grossmasstäblichen topographischen Karten. *Zeitsch. f. Vermessungswesen*, 82: 299-303.

Rathjens, C. 1976. Rezension Landformen im Kartenbild. Erdkunde, 30: 307-308.

Salistschew, K. A. 1967. Einführung in die Kartographie. 2 Bde, Gotha/Leipzig.

Sauškin, J. G. 1966. Die Entwicklungsperspektiven der sowjetischen Geographie. Aus der Praxis der sowjetischen Geographie. Gotha/Leipzig: 9-29.

Schmithüsen, J. 1973. Die Entwicklung der Landschaftsidee in der europäischen Malerei als Vorgeschichte des wissenschaftlichen Landschaftsbegriffs. *Geogr. Zeitschr., Beihefte*, 33: 70-80.

Schmithüsen, J. 1976. Allgemeine Synergetik. Grundlagen der Landschaftskunde. Lehrbuch der Allgemeinen Geographie, 12. Berlin – New York.

Schulz, G. 1969. Versuch einer optimalen geographischen Inhaltsgestaltung der topographischen Karte 1 : 25.000 am Beispiel eines Kartenausschnitts Blatt 6506 Reimsbach/Saarland). Berliner Geogr. Abhandl. 7.

Schulz, G. 1974. Die Atlaskartographie in Vergangenheit und Gegenwart und die darauf aufbauende Entwicklung eines neuen Weltatlas. Berliner Geogr. Abhandl. 20.

Werner, D. 1973. Interpretation von ökologischen Karten am Beispiel des Ätna. *Erdkunde* 27: 93-105.

Address of the author:

Prof. Dr. C. Rathjens,

Geographisches Institut der Universität des Saarlandes,

66 Saarbrücken, FRG

POTENTIAL ASSOCIATIONS OF CULTIVATED PLANTS IN TROPICAL CLIMATES

Ralph Jaetzold

Up to now, little attention has been paid to the relations of climate and cultivated plants by biogeographers, because they were more interested in natural vegetation. However, on account of the growing population pressure in developing countries, it becomes more and more necessary not only to know the climax vegetation but also the climax cultivation. This depends not only on climate of course, but climate is the first frame for any kind of land use pattern further differentiated by soils, marketing facilities and so one.

The first task is to create a climatic classification which is adequately differentiated in order to parallelize it with the climatic needs of the different cultivated plants and their varieties. This was tried by Papadakis (1966 and 75) and Schreiber (1973). But to deliminate climatic associations of cultivated plants, a more detailed classification was desirable. The author did this for the tropics (1970 and 1977a and b), where most of the developing countries are situated. This classification is given in a decimal system and sets up 50 main types and 400 subtypes for the tropics. Further differentiations are possible and are useful for certain purposes.

The climatic formulas show in the first letter the thermal zone, in the first figure the thermal type, and in the second the sub-type. A t stands for tropical. It means in the lowlands that the annual mean temperature is above 18°C and frosts do not occur. In the highlands frosts are possible, but the mean temperature (calculated for the sea level) has to be above 18°. This definition combines the ideas of H. v. Wissmann (1948) and C. Troll (1943) about the criteria of tropical climates.[1] The **thermotypes** are in the first order divided by the annual mean temperature, which gives a clear classification of the altitudinal belts (fig. 1), well known since A. v. Humboldt, although the critical values are slightly different. They are orientated to the limits of important cultivated plants. These limits have to be divided in a limit of economic cultivation and a limit of ecological existence, which are different in many cases. The economic limit depends also on the market situation and many other factors. Here, it is looked upon as a boundary where yields range below the average due to the climate. On the other hand, there exists a wide spectrum of varieties which are more or less adapted to marginal climatic conditions.

---

1 For certain purposes it is necessary to divide in **it** = inner tropics and **mt** = marginal tropics according to the reduced 20° C-isotherme of the coldest months. The potential associations described here, mainly refer to the inner tropics.

The annual mean temperature of 23°C is the critical value for the tropical hot climates t_1 (fig. 1). It is the upper economic limit of cola nuts, coriander and vanilla. It is also the lower limit for higher-quality tea and virginia tobacco, in semi-arid areas also for wheat (in humid areas it is to much endangered by rust).

The mean annual temperature of 17,5° within the absolute ground frost limit is the upper boundary of the tropically temperate climate t_2. In the equatorial zone it exists between 1800 and 2200 m and roughly corresponds to the upper economic limit for bananas[2], beans, sweet potatoes, to the upper ecologic limit for ananas, arabica-coffee, citrus fruits, cassava, mangoes, papayas, taro and to the lower economic limit for peas, potatoes and pyrethrum.

The mean annual temperature of 11°[3] with the possibility of frosts at all seasons is the upper boundary of the tropically cool climate t_3. From the economical point of view, it is the upper limit for barley, potatoes, carrots, pyrethrum and white cabbage in those innertropical countries, where the population pressure does not push up the limits for cultivation. Apart from that, cultivation still takes place beyond this boundary. Here, wheat is already touching the ecologic limit.

The annual isotherm of 2° as the limit of the tropically cold climate t_4 roughly corresponds to the upper limit for the existence of grazing plants (Walter 1970, p. 66 a.o.). Above this boundary comes a zone of bare frost-debris, which is covered with a few lichen species at the most. Therefore this boundary is also the upper limit for the pasturage of even unpretending animals.

It has turned out, that at least two subclasses of thermal zones are necessary respectively, for important vegetation and cultivation boundaries can also be found in the intermediate areas at 600 m, 1500 m, 2500 m and 3600 m.[4] As an upper limit for cultivated plants preferring heat and as a lower limit for those preferring coolness the 1500 m – boundary is particularly noticeable. Therefore, we distinguish in the second decimal (fig. 1)

1 warmer subclass

2 cooler subclass

The treshold values are determined by the minimum temperatures, because a number of tropical plants are very sensitive to coldness. This not only concerns temperatures below zero but also higher minima up to 18° can be limiting by influencing protein synthesis. In order to be in conformity with as many cultivated plants as possible, the threshold values for the thermic subtypes of the first order were chosen accordingly:

2 Rural growing of bananas, for commercial purposes and for cooking bananas, the limit is at an annual mean temperature of about 20°.

3 The boundary is not identical with the classical upper limit of the tierra fria which depends on the line beyond which forest no longer grows and may be between 12° and 8° according to the species of trees (Andrade Marin 1945; Lauer 1975, p. 320).

4 The data given in meters are only approximate values. The amount of radiation is lower windward due to the upwind-clouds of the trade wind and therefore the altitudinal limits are up to 200 m lower. On the highly radiated lee sides though, they can be up to 200 m higher. In big massives the limits are higher than in small ones.

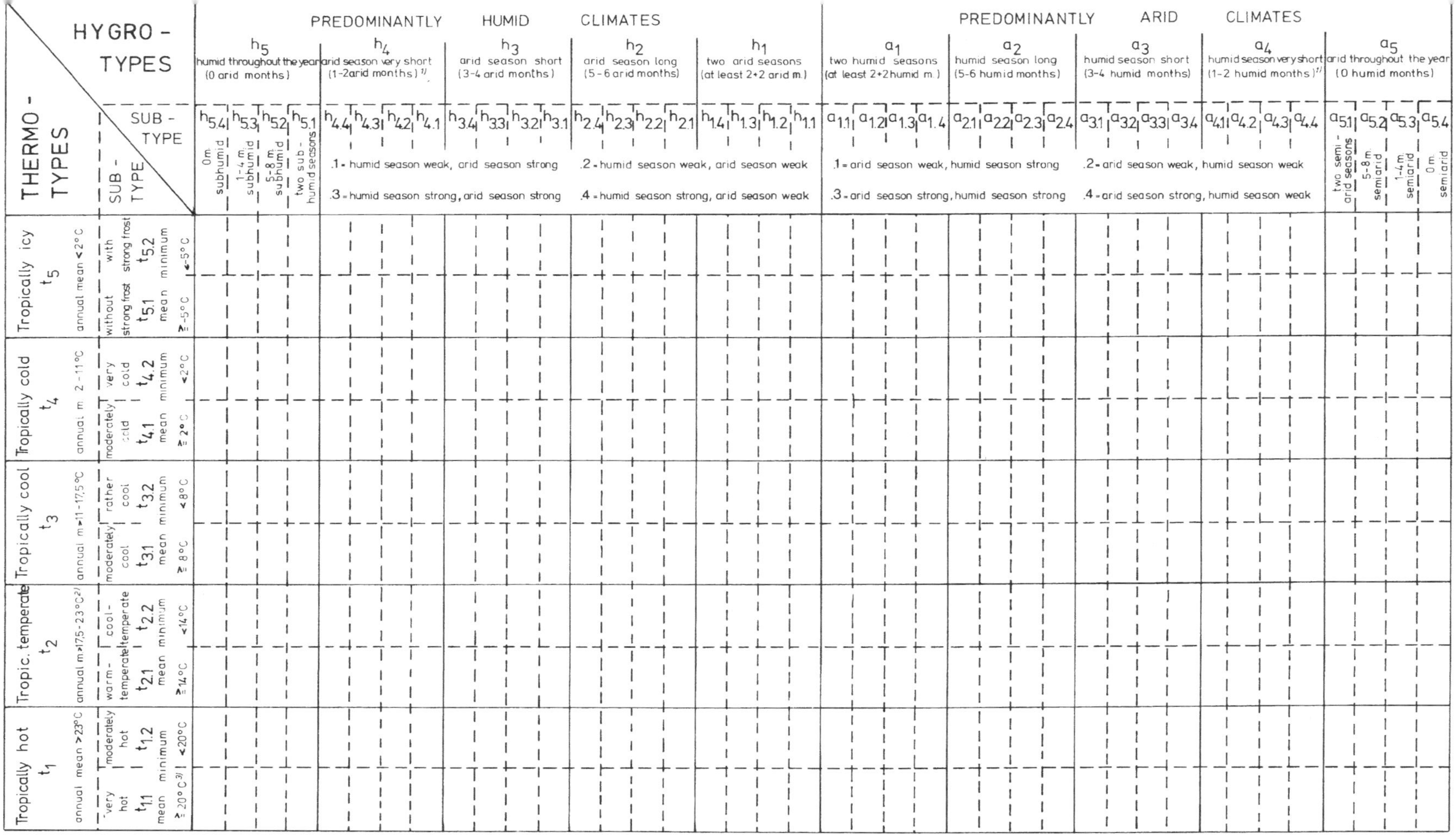

Fig. 1. The tropical climates in a decimal system

1) Also 1+1, 2+1 or 1+2 months.　　2) Absolutely free of ground frost.　　3) Mean minimum of the coldest month > 18°C.

Thermo-types	Th.sub typ. \ H-sub types (Hygro-types)	h_5	h_4	h_3	h_2	h_1	a_1 [1)]	a_2	a_3
		h h h h 5.4 5.3 5.2 5.1	h h h h 4.4 4.3 4.2 4.1	h h h h 3.4 3.3 3.2 3.1	h h h h 2.4 2.3 2.2 2.1	h h h h 1.4 1.3 1.2 1.1	a a a a 1.1 1.2 1.3 1.4	a a a a 2.1 2.2 2.3 2.4	a a a a 3.1 3.2 3.3 3.4
t_3	$t_{3.2}$	p (po) (w) (by) (py)	p by (po) (w) py	(p) by (po) py(w)	(p) by (w) p (py) po (po)	py p by (by py) po (w) po	(p by po w)	by (w) p (p) (po)	(p) (by)
	$t_{3.1}$	p (po sp t) (w) (by m py)	p bv (po m) (sp) t w (py) py	p bv po (m) p (sp) w (t)	(b) p b (sp) w po (m) pv (po py	p by (po) (w) (m) (m) py (t) (p by py)	(by) p (p) (po w)	by po w p (m) (p) p (p)	(by) p (po) (po)
t_2	$t_{2.2}$	ta (ta) (b ba) (m py) t (sp) t	b (ba m) sp t (ac) ac (ac py) ta (ta)	b ba m (py) sp (ac) ac t (t) t (w)	b (by so) w ba (ac) si (t) (ba) m sp (m)	(si so) (w) ac (ac) m (t) m (by) (ba sp) by cr (cr)	(m) (b) (si) (so) (w) (w) (by cr sp)	by (so) w m m sp (m) (sp) b cr (b)	cr (so) (by) sp (w) cr cr (b) (w)
	$t_{2.1}$	(b cs t) ta ba (m) rc (sc)	b ba m cs sp (so) ac (pe) t sc co (r) sc ta (rc) (ct)	b ba (co) m co se (t) ta sp (pe) (so) (ac rc r) (ct) (t sc) si	b co mb (se) so (pp) ba vt (ba) ct (si) (ac) vt (pe by) m (r) si sp (m sp)	(ac) b (by co (cs) pm cr si so m (m) pp ba (ct pe se) sp (sp)	(co mb pm pp) b b (m) (b) (m) si (pe by cs cr so sp) (b si)	b (pm) (se) so pp ct sp (ct sp) (pe) si (w) si (by si) m (cs) cr vt (m w)	(by co mb pm) (m si) b (pe) (so sp) (b cr) (m) cr cr (pe w) (sp so)
t_1	$t_{1.2}$	(cs r sk) ta (sc) ba (m) rc	(b) ba (cp cs) (op) r (so) (pe) sp ta rc (t) (sc) ct sc (po)	b ba po m cs r se sp (pe rc) so rc (ta) ct (op) (sc) si	ba co cs se rc (ta) so pp ba (ba) ct (pe) si m r sp (m sp)	(co cs) pm cr si ct (ba se ta pe m ct pe m (b) ba se sp (sp)	(co mb) pm (ct m) (pp) (se) so (so) (b pe cs sp) cr si (si)	(b) co pm se so pp pe (pe) pe mb si (mb si) ct m (cs) cr sk (ct pe)	(b by co mb) pe pe(m) pm (m si) (so) (se sp) (pe) (se sp) (so) cr (cr) cr (cr)
	$t_{1.1}$	(cp) ca (cs so) (r) ta (sc) ba ca (m) (sp)	ba ca cp ct (m) cs (so) op rc sp y ta (rc) (sc) (cn co se) sc	(co) cn (co) cp (m) y (cs) r sp (se) so (cc ta) (pe) (ba) ba cc op sc si	cn co cs se so (pp) r sp (sp cp) ba (ta) (ba) ct (pe) si m (m)	cn (co cs) pm cr r so (ba se si (ct) pe (m) cp pp ba se sp (se sp)	(co mb) pm si se (se) (pp) (ct m cp) cr se (pe cs sp) cr so (si)	co pm se so (pe) pe (pe) pe cn si mb (cn si mb) ct m (cs) cr sp (m)	(co mb) pm pe (pe) pe(m) (m si) (cr) cr (so se (se so) (so) cr (sp) (cr)

1) When there is 2 + 2 humid months, only special varieties of cultivated plants

( ) = less favourable climate

ac	arabica coffee	cc	caoutchouc	cs	cassava	p	peas	py	pyrethrum	si	sisal	vt	virginia tobacco
b	beans	cn	cashew nuts	ct	cotton	pe	peanuts=groundnuts	r	rice	so	sorghum	w	wheat
ba	bananas	co	cowpeas	m	maize	pm	pennisetum millet = bulrush m.	rc	robusta coffee	sp	sweet potatoes	y	yute
by	barley	cp	coconut palms	mb	mungo beans	po	potatoes	se	sesame	t	tea		
ca	cacao	cr	castor	op	oil palms	pp	pigeon peas	sc	sugar cane	ta	taro		

Fig. 2. Potential associations of cultivated plants in the agro-climates of the inner tropics

For the boundary between $t_{1.1}$ and $t_{1.2}$ the mean minimum of the coldest month − being 18° − was chosen[5], because many typical permanent cultures of the tropical low-lands react on cooler nights by delayed growing. The value corresponds to the upper economic limit for cashew-nuts, cloves, jute, cocoa, coconuts, oil palms, rubber (Hevea brasiliensis), yams, and at the same time, it approximately characterizes the lower economic limit for anise, oriental tobacco and robusta coffee.

The mean annual minimum of 14° = boundary between $t_{2.1}$ and $t_{2.2}$ approximately forms the upper economic limit for aubergines, cotton, exportable citrus fruits, groundnuts,[6] cassava,[6] cooking bananas[6], melons, bulrush millet, robusta coffee, sesame,[6] virginia tobacco and sugar cane. In addition to that, it is the upper ecologic limit for jute and the lower economic limit for barley and in humid areas also for wheat.[6]

The mean annual minimum of 8° = boundary between $t_{3.1}$ and $t_{3.2}$ is important as a limit for finger millet and maize (Nuttonson 1948). For sorghum and sweet potatoes it is generally too cold.

The mean annual minimum of 2° (resp. 6 months frosts not below −3° and mean maximum above 12°) = boundary between $t_{4.1}$ and $t_{4.2}$ forms the absolute upper limit for bitter potatoes, barley, carrots, oca (*oxalis tuberosa*) and white cabbage.

Of course also maximum temperatures and variation widths have an influence upon land use, so that further divisions result therefrom. This, however, would bring up to much details here.

The **hygric classification** has been developed on the basis of the duration and intensity of humid and arid seasons as the most important factors of climate for agriculture. The first decimal indicates the duration and distribution of the humid and arid seasons respectively, while the second indicates their intensity. A preceeding 'a' of 'h' will tell whether there is, according to the year's water balance, a predominant arid or humid climate.

h_1 = predominantly humid climate with two arid seasons (about 5-7 humid months separated by more than one arid month)

h_2 = predominantly humid climate with a long arid season (6-7 humid and 5-6 arid months, when precipitation is high, 5 humid months are sufficient)

h_3 = pred. humid climate with a short arid season (8-9 humid and 3-4 arid months)

h_4 = pred. humid climate with a very short arid season (10-11 humid and 1-2 arid months)[7]

---

5 In order to keep the system of climatic types uniform, the mean annual minimum of 20° was used in fig. 1. With regard to the typical monthly-mean variation of 4° per year in the inner tropics this largely corresponds to 18° of the coldest month.

6 As far as this line is identical with the 20° annual isotherm, which is mostly the case in the inner tropics.

7 In equatorial latitudes, there can be an additional arid month to signify the second very short dry season.

h_5 = humid climate without an arid season (12 humid months)

For the **subdivision** there are four subtypes, according to intensity:

1. The humid seasons are weakly, the arid seasons are strongly developed.
2. Both are weakly developed.
3. Both are strongly developed.
4. The humid seasons are strongly, the arid seasons are weakly developed.

For the humid part, weak = sub-humid, strong = full-humid. Sub-humid means that precipitation = 0.4-0.8 of the evaporation E_0 of an open water surface; full-humid: precipitation $>$ 0.8 of evaporation (this value corresponds with the **average evapotranspiration** E_t of most cultivated plants, that is, it could be said: precipitation = 0.5-1.0 E_t is sub-humid, precipitation $>$ E_t is full-humid). Most cultivated plants start with an evapotranspiration of 0.4 E_0, reaching 0.8-1.2 E_0 in the main growing period (if there is enough water), and during ripening stage they use 0.4-0.8 E_0 The arid part is divided in the middle in order to distinguish strong = full-arid from weak = semi-arid, that means the critical value is 0.2 E_0 or 0.25 E_t respectively (for many cultivated plants this comes close to the fading point).

$h_{1.1}$ = humid months predominantly sub-humid, arid pred. full-arid
$h_{1.2}$ = humid months predominantly sub-humid, arid pred. semi-arid
$h_{1.3}$ = humid months predominantly full-humid, arid pred. full-arid
$h_{1.4}$ = humid months predominantly full-humid, arid pred. semi-arid
$h_{2.1}$ etc.

In the h_5-climate, there are no arid months. However, seasons with light precipitation which are only sub-humid may occur. Thus, the subdivision is modified:

$h_{5.1}$ = nearly all-year humid with two sub-humid seasons
$h_{5.2}$ = nearly all-year humid with one extended sub-humid season
$h_{5.3}$ = nearly all-year humid with one short sub-humid season
$h_{5.4}$ = nearly all-year full-humid

The **arid climates** are divided like the humid ones

a_1 = predominantly arid climate with two humid seasons (about 4-6 humid months which are separated by more than one arid month)

$a_{1.1}$ = arid months pred. semi-arid, humid pred. full-humid
$a_{1.2}$ = arid months pred. semi-arid, humid pred. sub-humid
$a_{1.3}$ = arid months pred. full-arid, humid pred. full-humid
$a_{1.4}$ = arid months pred. full-arid, humid pred. sub-humid

a_2 = predominantly arid climate with an extended humid season (5-6 humid and 6-7 arid months)

$a_{2.1}$ = arid months pred. semi-arid, humid pred. full-humid
$a_{2.2}$ = etc.

a_3 = pred. arid climate with a short humid season (3-4 humid and 8-9 arid months)

a_4 = pred. arid climate with a very short humid season (1-2 humid and 10-11 arid months)[8]

a_5 = arid climate without a humid season (12 arid months)

There are normally no humid months. However, seasons with higher precipitation being only semi-arid may occur, which is sufficient for some semi-desert plants, so that a periodical pasture develops. Thus, the division is modified:

$a_{5.1}$ = nearly all-year arid with two semi-arid seasons

$a_{5.2}$ = nearly all-year arid with one extended semi-arid season

$a_{5.3}$ = nearly all-year arid with one short semi-arid season

$a_{5.4}$ = nearly all-year full-arid

In spite of the system-like character of the threshold values, they do have distinct relations to the distribution of cultivated plants:

h_1 = Permanent cultures, insensitive to several short dry seasons (bananas, for example) and only short-lived annual cultures are possible, since the humid seasons are only short, too.

h_2 = Permanent cultures resisting long dry seasons (i.e. arabica coffee) and almost all annual cultures are possible except for hygrophiles.

h_3 = Permanent cultures resisting a short dry season (i.e. oil palms) and nearly all annual cultures are possible.

h_4 = All permanent and annual cultures are possible except for those preferring dryness.

h_5 = All permanent and annual cultures are possible except for plants which do not like wetness at certain times (i.e. when maturing) but great danger of fungus diseases.

a_1 = Permanent cultures resisting dryness (i.e. sisal) are still possible, annual cultures are too much endangered by dryness, since both of the humid seasons are very short.

a_2 = Annual cultures with a short or middle-term vegetation period are well possible, since the humid season lasts relatively long. Also, annual cultures resisting dryness can be planted.

a_3 = Short-lived annual cultures are possible. Too dry for permanent cultures.

a_4 = No rain-fed cultivation is possible, it is a climate for ranching.

a_5 = Only nomadic animal husbandry is possible.

For the subtypes, there are many various relations. Therefore, they can only be looked up in the table (fig. 2).

The combination of the thermal and the hygric differentiation produces 50 climatic types and 400 subtypes. Each one of the climates suitable for cultivation has its own potential association of cultivated plants (fig. 2).[9] Agronomists in developing countries need maps with this system in the International World Map Scale of 1:1 Million (Jaetzold 1977 b) or at least 1:2.5 Mill. to pick up ideas for the agricultural extension service. Such maps are existing for East Africa (Jaetzold

---

8 In equatorial latitudes, there can be an additional humid month to signify the second very short wet season.

9 This applies to normal sandy loam soils.

1977 a and b) and will be published for Northern Peru and the Senegal. Specialists of other countries are invited to join the programme.

For certain cultivated plants, there are favourable climates of the first order if the thermic and hygric conditions are favourable. They are of the second order if one of the two climatic factors is less favourable (such cases are in brackets in fig. 2).

In spite of the detailed climatic differentiation, there are still climatic limits for cultivated plants, which exist in between the threshold values of the classification. The lower part of $t_{2.1}$, for example, is already too warm for arabica-coffee. However, the system could not be made more complex. At a first glance, it is rather confusing anyway, but this degree of specification is necessary, in order to look up the potential associations of cultivated plants of each tropical climate.

In dry areas it is useful to include the variability of rainfall. This can be done in a third decimal, which shows the effective rainfall, that is, precipitation that promotes growth, during the vegetative period in a 2/3 probability. [10] This is expressed in numbers of which each number indicates uprounded 100 mm of the minimum effective rainfall in two out of three years on the average (2 = 200 mm, 3 = 300 mm and so on). If the amount is spread over two rainy seasons, two numbers are used, for example 5 + 3, which means that during the long rains there are at least 500 mm and during the short rains 300 mm in two years out of three.

The system will be extended in other climatic zones. There the soil moisture remaining from a cold period has to be added to the effective rainfall. Critique and cooperation are welcome.

References

Andrade Marin, L. 1945. Cuadro Sinoptico de Climatologia Ecuatoriana. Quito.
Humboldt, A. v. 1814-25. Voyage aux régions équinoxiales du Nouveau Continent fait en 1799-1804. 3 Vol. Paris.
Jaetzold, R. 1970. Ein Beitrag zur Klassifikation des Agrarklimas der Tropen. Festschrift für Herbert Wilhelmy. Tübinger Geogr. Studien 34: 57-69.
Jaetzold, R. 1977a. Das Klima. In: W. Leifer (ed.), Kenia: 30-50.
Jaetzold, R. 1977b. Klimageographie Blatt Lake Victoria 1:1 Mill. des Afrika-Kartenwerkes. Berlin.
Lauer, W. 1975. Vom Wesen der Tropen: Klimaökologische Studien zum Inhalt und zur Abgrenzung eines irdischen Landschaftsgürtels. Wiesbaden. (Abhandlungen der Mathematisch-Naturwissenschaftlichen Klasse, 3.).
Nuttonson, M.Y. 1948. Some preliminary observations of phenological data as a tool in the study of photoperiodic and thermal requirements. *Chronica Botanica comp.* Waltham, Mass.
Papadakis, J. 1966. Crop Ecologic Survey in West Africa. Rom.

10 Precipitation is considered to be effective when it occurs during an agrohumid period, deducting the amount of runoff and ineffective precipitation. Agrohumid means that precipitation and soil moisture are generally sufficient for the growth of crops.

Papadakis, J. 1975. Climates of the World and their Potentialities. New ed. Buenos Aires.
Schreiber, D. 1973. Entwurf einer Klimaeinteilung für landwirtschaftliche Belange. Paderborn.
Troll, C. 1943. Thermische Klimatypen der Erde. *Pet. Geogr. Mitt.*, 89: 81-89.
Walter, H. 1970. Vegetationszonen und Klima. Stuttgart.
Wissmann, H. v. 1948. Pflanzenklimatische Grenzen der warmen Tropen. *Erdkunde* 2:81-92.

Adress of the Author:
Prof. Dr. Ralph Jätzold
Geographisches Institut der Universität, Trier

PRINCIPLES AND MODELS AS TOOLS FOR ECOSYSTEM-RESEARCH, WITH EXAMPLES FROM THE AMAZON BASIN

Harald Sioli

Abstract

Model-building, one of modern methods in ecosystems research, is confronted with the elaboration of 'principles' which determine certain characteristic features of ecosystems and rule their functioning. Advantages and disadvantages as well as the limits of both methods are discussed. Research on the forest-covered Amazonian lowland where 4 dominating principles could be discovered and established, is taken as an example for demonstrating the validity of both methodical procedures and of the possibility of predictions to be made by their application on consequences of interferences in those ecosystems.

Modeling has become, in more or less the last decade, one of the predilected tools in modern ecosystems-research. It is being praised and recommended many times as the only reliable key for the understanding of those complex and complicate dynamic structures which are many, if not most ecosystems on earth, and accepted and performed especially by those ecologists who realize in nature predominantly its cybernetic aspect, neglecting in the same time other, just truly biological characters in the internal context of ecosystems as there are e.g. the morphological problems (summarily even understood only as 'counting bristles on some pet species' (Jansson, 1976)), phylogenetic, physiological, ethological, psychological ones, etc. On the other hand, thorough critics on the unlimited validity of such model-making have been published (e.g. Hedgpeth, 1977).

I don't want to enter, too, into the discussion about modeling since I think that the pro- and contra-standpoints are linked with the extents of the biological horizons of the authors. Instead, I wish to demonstrate how far the perception and knowledge of some basic and characteristic principles in e.g. a selected large landscape-ecosystem may lead us to an understanding of its functioning, and also how far the construction of models may be helpful in that effort and present us with new insights. Few points of the research on the ecosystem of the forest-covered Amazonian lowlands will be cited as an example.

The ideal and ultimate goal of a perfect model is a quantitative, mathematical one in which all facts and factors have been quantified and related to each other in form of a mathematical formula. Therefore, no unforeseeable events are theoretically admitted in such ideal perfect models of ecosystems so that all consequences of contacts with other ecosystems or outside areas, or of human interferences, become exactly predictable. And when that predictability is not achieved by a constructed model, then that model is considered as not yet perfect, that some data are still missing and more research on more facts and factors must be carried out.

145

The elaboration of principles — as common conditions in an ecosystem to be studied, being them some important basic environmental facts which determine certain characteristics of the ecosystem, or being them some structural or functional general features of the ecosystem as necessary adaptations to certain basic environmental facts —, however, recognizes ab initio our limited knowledge and doesn't tend to envisage absolute rigid quantitative parameters. It means that conclusions are drawn from the results of as many studies as possible made in as many places and periods as possible within a given regional ecosystem. The comparison of these results may reveal a uniformity of a large ecosystem in certain respects, or also show differences which then may lead to sub-divisions of the large ecosystem into smaller ones, each with certain common principles of its own.

One characteristic of such a principle is its 'Grenzunschärfe' ('unsharp borderline'). Schwabe & Klinge (1960) have pointed out 'Grenzunschärfe' as being typical for the biosphere since it makes part of the character of life itself. In our case, 'Grenzunschärfe' of a stated principle is related to the fact that no ecosystem at all is limited by exact and definite borderlines, neither in space nor in time: Ecosystems are open systems. There are always transition zones of varying extensions, and the ecosystems also change, by external and internal reasons, over shorter and/or longer periods of time. For to cite only two examples of instability in time (the instability in space is obvious): Climatic changes — the reasons of which are not predictable (cf. Gerwin, 1978) — may alter an ecosystem profoundly, and the new ecosystem may for its part influence the climate again; or retrograde erosion may connect different river-systems with different aquatic biotas so that foreign species of plants or animals may immigrate and alter the biotic structure and therewith the functioning of the ecosystem, etc.

Principles, thus, are abstractions found realized in nature in greater or lesser degree within ecosystems, and the realizations are, under certain aspect, comparable with a 'tema con variazioni' in music. I.e. there is a vast field for variations in the stamping of ecosystems under the dominating influence of one and the same principle — which however is never the only one to rule an ecosystem.

A principle, once ascertained and established, is thus no good for predicting the detailed and quantitative consequences of changes inflicted to the ecosystem, but it does allow to predict that, when the actual principle is being altered or destroyed and substituted by another one, the ecosystem will correspondingly change into a direction determined and ruled by the new principle, again with its proper character and 'Grenzunschärfen'.

Before passing to examples from the forest covered Amazonian lowland, let us consider still the question whether it would really be desirable for mankind to be able to exactly and quantitatively predict all future events and changes which ecosystems on earth have to undergo by natural or — what is actually acute — human interferences. That we are in fact unable to such concrete predictions of future situations becomes clear when we consider that all ecosystems are open ones, susceptible for unpredicted influences from outside, and that the events on earth depend even on cosmic influences, or that we really do not understand the causal laws of organismic speciation, for only to cite two aspects of that unability of ours.

But the desirability of concrete prediction is another question. In former publications (Sioli, 1972, 1973a) I have tried to demonstrate that surprise is one of the indispensable requisites for the 'play of life' which, without surprises, would come to an end. Accepting the desirability of concrete quantitative predictions without chances of errors or of 'Grenzunschärfen', i.e. without surprises, would mean that our aim would be a world reduced to a mechanism with man and all organisms as prefabricated wheels in it, without any fate or subjective experiences ('Erlebnisse'). Succeeding in the theoretical goal of modern computerized modeling of the life-processes on earth, as they are bound to ecosystems, would finally bring the fulfillment of the sinister prophetic words of Richard Benz (1923) in one of their points: 'Die Vergnügungsmaschine, die Rechenmaschine, die Kriegsmaschine vernichten schließlich die Menschheit' ('The amusement-machine, the calculation-machine, the war-machine finally destroy mankind').

All this, however, doesn't mean that modeling in se be evil and compulsorily must lead to destruction. That is the case when the sense and the possibilities of model-making are not understood and the results of such misinterpreted modeling be applied to our life through its environment. Isn't behind of that attempt the mania for power, diverting natural research definitely from its most noble aim of cognition?

Sense and value of modeling lie in another direction and are comparable to some degree to the procedure of an architect who wants to build a good house. First he conceives the idea of that house, establishes its principle or principles, and then develops the external and internal morphic and functional structure. And finally he produces a designed or even a threedimensional 'model' of the house, a 'maquette' which represents the house also in its surroundings. The model serves for him, or his orderer, to test his basic idea now in all its concrete and quantified details whether all previously envisaged parts fit harmoniously together and are functionally correct, and how the whole house structure fits into its environment.

We see, such a model is a more or less complete transfer of just **one certain** house into more easily understandable material and dimension, and never a valid expression of the house.

The same is true for modeling of ecosystems: a model may represent more or less exactly the functional interrelations within one peculiar ecosystem which in reality exists as a singular individuum (not even twice in form of true 'identical twins'), or within one peculiar group of several ecosystems linked together — as open systems! — to one greater, comprising ecosystem as a higher unit. It is never representative for **the** ecosystem of a certain type, be it **the** forest-ecosystem, or **the** lake-ecosystem, etc.

Only one case might be imaginable in which a model would be valid for all ecosystems on earth together, namely when it represents the greatest ecosystem we can idealize: that of our whole globe. For that is, as far as we can look into the cosmos, also an individual singularity. But who would dare to build such a model? And if some super-genius someday would achieve to come near to that ultimate goal, would that overwhelmingly complex model not be as unintelligible as the reality of its global subject itself?

All other exact and quantified models, as already said, are not representative for the ecosystem of a certain type: the forest-ecosystem, the lake-ecosystem etc. is an abstraction which we find not to exist in nature when we look for structural and quantitative details, thus comparable to Goethe's 'Idea of the Plant'. It is, however, a true category in our thinking processes which results from the fact that several different real ecosystems have some principle, or principles, in common.

Therefore, it is not permissible to conclude from a model built on one peculiar individual ecosystem — and only by that way a model in modern scientific sense can be built — on other ecosystems even if of the same type.

It is permissible, however, to assume that in different individual ecosystems which have a certain principle, or certain principles, in common, there exist certain structural and/or functional similarities, namely in those reaches which are ruled by these principles. And then the procedure of model-building of concrete individual ecosystems belonging to the type in question may become very useful for more easily to understand the 'tema con variazioni' and to test the limits of validity of certain selected principles.

But also between principles and models there may exist 'Grenzunschärfen', namely when a principle enters the dimension of always more detailed quantifications, or when a model is not yet concrete enough for exact quantification so that it resembles more a principle. In such a case it should better be presented as a working hypothesis instead of a model.

Let us try now to cite few selected examples from the Brazilian part of the Amazonian lowland with its regional water bodies and with its forest.

The first scientist who ever recognized and outlined the most striking basic ecological characteristics — or principles — of the Amazonian lowlands was Hans Bluntschli who stated: '... that **wind** and **plain**, **forest** and **water** act intrinsically together, and we perceive that all and everything must stand under their influence in Amazonia, from the smallest living being to the activity and the behaviour of mankind ...' (Bluntschli, 1921, p. 51; translated by the author).

We shall concentrate on forest and water. Bluntschli who was a comparative anatomist observed already with his eyes schooled for structural interrelations, that, while forest and water 'act intrinsically together', they depend on each other, thus being mutual expressions of each other. In his statement he visualized first of all the circulation of water through the Amazonian landscape-ecosystem (which shall be treated more below). He didn't make any chemical water-analysis and evidently also didn't know, or consider — for he was an eye-man! —, the first and for a long time only data of such analyses of Amazonian river-, creek- and groundwater made by Katzer near the end of last century (Katzer, 1903) so that he couldn't yet know anything about their chemical conditions and the relationships between these ones and the forest via the chemical characteristics of the soils. If he had known he certainly would have linked forest and water not only by the steady availability of the circulating liquid 'water' alone but also by the chemistry of the natural water bodies since groundwater, creeks and rivers represent extracts of the soils on which the forest stands.

Katzer (l.c.) had first discovered the surprising chemical pureness and poorness of most of Amazonian waters. When, starting after the second world war, systematic research on Amazonian water chemistry was undertaken and pursued till today, that discovery by Katzer has been confirmed and extended to the by far largest areas of the whole forest-covered Amazonian region, the 'hylaea' of Alexander von Humboldt (Sioli 1950, 1954, 1955; Schmidt 1972; Furch 1976).

Exceptions are only few small local zones because of geological-lithological reasons (Sioli 1963) and especially the pre-andean strip of land with its projections along the white-water-rivers with the alluvial lands of their 'várzeas'. Here we find chemically rich waters, to be taken as signs for also richer, more fertile soils. These zones are occupied by a different forest-vegetation of many species restricted to them and which do nor occur on the 'poor' lands. The ecological sub-division of Amazonia which Fittkau (1971) established on such geochemical basis also shows very strong 'Grenzunschärfen' which are not only due to lack of chemical data in many and great extensions of the region but also typical for ecosystem-boundaries.

Most of the Amazonian waters yet revealed themselves to be so pure and poor chemically that they are best compared with distilled water of low quality and equal almost to rainwater (see Fig. 1).

In a humid climate like that of the Amazonian hylaea, and under a forest cover in a climax state, i.e. with a biomass constant over a long period of time, all soluble substances which are liberated in the soil by weathering processes, are being washed out and, since they are not additionally accumulated in an increasing amount of vegetation, must appear in the ground-, spring- and creek-waters. And when these are chemically so pure and poor as described, we must conclude that the soil is correspondingly poor and does not contain any appreciable quantities of reserves of substances, among them the nutrients for plant growth, which could be mobilized by weathering. That implicitly means that those soils are extremely unfertile.

In spite of this fact we find an exuberant forest standing on those soils. That seems, at a first glance, a paradox, and also Humboldt, impressed by the overwhelming exuberance of the Amazonian high forest, had already concluded that such a luxurious vegetation must grow on a most fertile soil!

The solution of that paradox, however, is that that forest does not grow **from** that soil but that it only stands **on** it, using the soil not as a source of nutrients but almost only as a substratum for mechanical fixation while living in a tightly closed circulation of the nutrients within the living matter.

We understand that chemical analyses of natural waters, mostly later also of soils (Camargo 1949, Sombroek 1966, Falesi 1972), have provided us with the comprehension of one fact or principle which is basic for the forest-ecosystem of the by far greatest part of the Amazonian hylaea: the poorness in nutrients of its

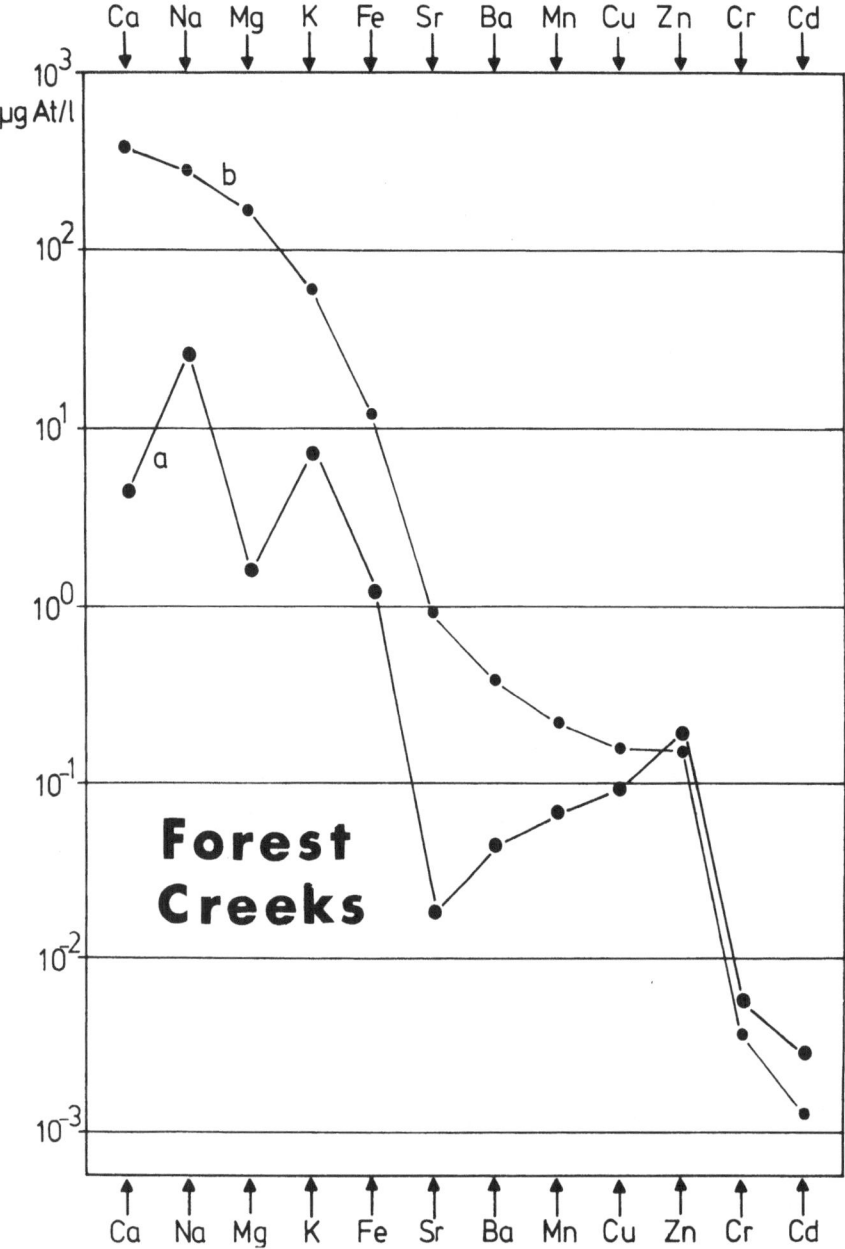

Fig. 1. Average values of chemical analyses of 10 central-Amazonian forest creeks (left side, a) compared with those of rainwater of Central Amazonia (right side, a). The values b on each side represent the world-averages of freshwaters. From: Furch 1976.

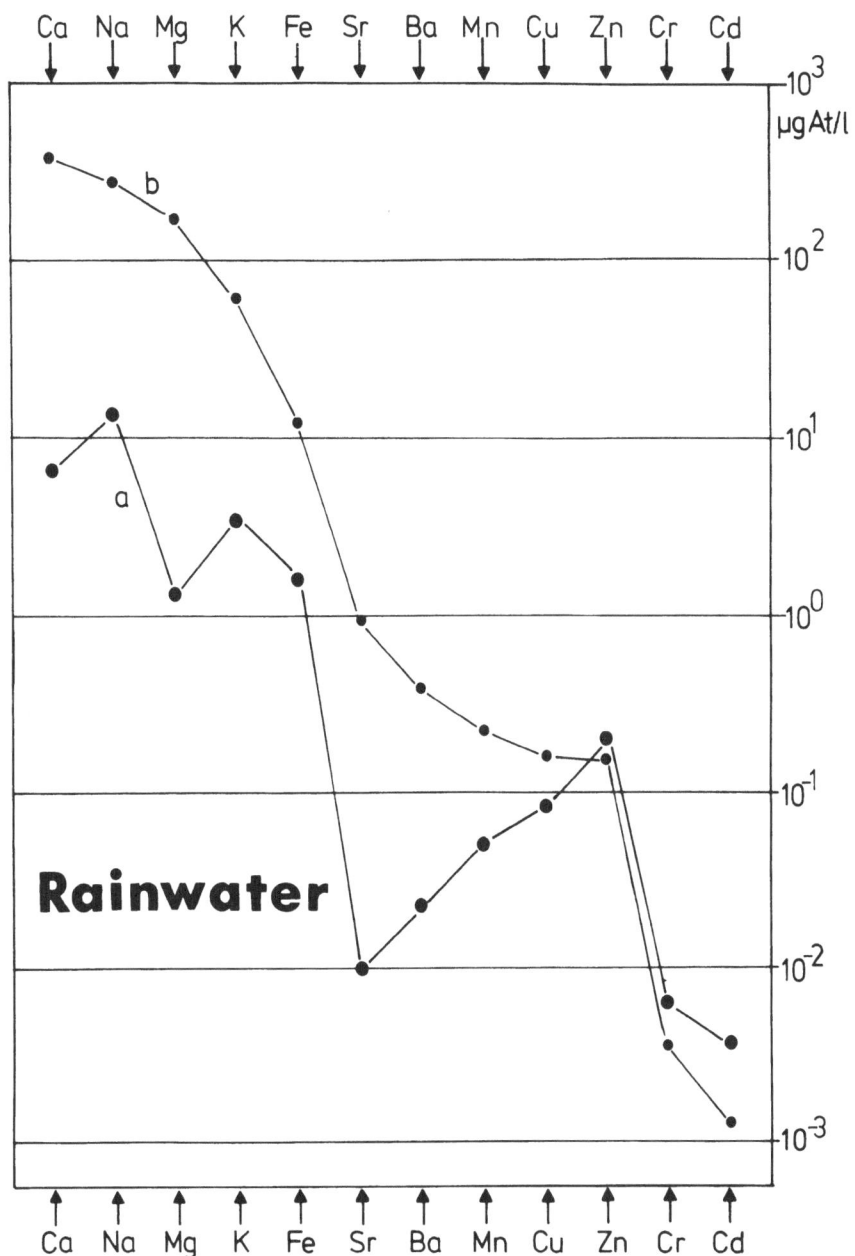

Rainwater

soils or, in plain words, the infertility of them. The answer of the living side of that forest-ecosystem to that challenge by a basic environmental fact then must also be taken as a principle in the functioning of the system: the closed circulation of the nutrients within the living biomass.

That principle, however, does not yet explain how the forest manages to realize the closed circulation of nutrients. When we look nearer to it and study it in more details, we find that nature has applied all possible means and methods to maintain the circulation closed, to reduce losses of nutrients by leaks in the circuit to a minimum. There is no appreciably thick humus layer on the Amazonian forest soil. Instead, the rootsystem of the forest trees is shallow, usually concentrated in the uppermost soil-layer 20-30 cm thick, and in average three times as dense as that of a temperate forest (Klinge 1973a). That indicates that the nutrients are never temporarily stored in dead matter on top of the soil (from which they might easily be washed out by the high rainfall) but are soon re-absorbed from the forest litter which rapidly decomposes under the wet-hot climatic conditions. Stark (1969) discovered first that decomposition of the dead forest litter – and therewith the recycling of the nutrients within the forest system – is being performed by fungi. And recently Harrera et al. (1978) were able to show fungi with one end of their hyphe sticking inside a decomposing litter leaf, and with the other end inside a living root and transporting [35] P directly from the litter into the root.

How closed the nutrient cycling in the living Amazonian forest is, could also purely chemically be shown by W. Franken (verbal communication) who, in a protected forest-reserve area East of Manaus (Central Amazonia), analyzed rainwater, the water dropping from the forest canopy, stem-runoff and groundwater: While the rainwater was chemically as poor as to be expected in natural areas far from industries etc., the dropping water and the stem-runoff were very rich in dissolved matter, among them the nutrients, – interestingly varying in their compositions sometimes strongly between different tree species – but the groundwater was practically as poor again as the rainwater. That means that the dense rootnet of the forest trees acts as a highly effective filter which retains all the dissolved substances offered by the dropping water and the stem-runoff, conducting them immediately back into the living forest trees without any losses into the groundwater, creeks and rivers by which they would be taken finally into the ocean and definitely lost for the landscape with its forest-ecosystem. (Cf. also Klinge & Fittkau 1972.)

Focussing our attention now more to the structure of the biota of the Amazonian forest-ecosystem, we see that, once overcome the basic difficulty for perpetuation of it: the deficiency of continuous delivery of fresh nutrients, nature has here developed the most diverse ecosystem we know on earth (cf. Fittkau, 1973) with the greatest known number of plant and animal species, by splitting the big general cycle of nutrients into an immense number of sub-cycles (or vice-versa: by the now given opportunity for an immense number of species of organisms to evolve, the big general cycle has been split into those very many sub-cycles, both processes certainly being connected by feedbacks). This diversity may also be taken as another 'principle' of Amazonian forest-ecosystem.

Does such 'diversity' indeed mean stability of the ecosystem as has been postulated by a number of ecologists? The answer cannot be easy and unequivocal since 'stability' is only a relative concept, especially against human interferences. Quantitatively and qualitatively different interferences may bring consequences of different quantitative and qualitative kinds and may have very different effects on the 'stability' of the ecosystem.

The diversity of the Amazonian forest-ecosystem with its enormous number of plant and animal species doesn't only mean that there is an equally enormous number of ecological niches occupied and utilized by those species, or of sub-cycles of nutrients intercalated in the closed big one. It also means that these plant and animal species together with the abiotic environmental conditions form, by their mutual interactions, a complexity with a perhaps still greater number of homoeostatic circles, from tiniest ones to big ones, which all act together; and simultaneously they all depend on the circumstance that the greatest homoeostatic circle of that system, that of the closed matter circulation, which results on that cooperation of the smaller ones, functions in a way of the least losses. The whole circulation system may be compared, cum grano salis, with the scheme of the blood circulation in an organism in which there are blood vessels of the most diverse sizes, from capillaries to the main arteries.

Let us give a concrete example which relates to the closed general matter circulation in the forest as proved by Franken's chemical studies. Part of the nutrients drained from the canopy with the dropping and stem-runoff waters is being extracted by the rainwater directly from the leaves and thus directly returns into the living tree. But most of animal life in the Amazonian forest happens in the canopy where there lives all that rich fauna of climbers and flyers as monkeys, coatis, sloths, birds, the great army of insects, etc. etc. These animals live directly or indirectly from the organic matter of the trees, and the endproducts of their metabolism, faeces and urine, are then being washed down by the rainwater, and their soluble contents will appear in the dropping and stem-runoff waters. On the way down the leaves, branches and stems of the forest trees, the enriched water may also come into contact with the epiphyllic and epiphytic plants supplying them with nutrients and establishing another more sub-cycle of nutrients within the big general one.

Trying to analyse and to grasp qualitatively and quantitatively all those sub-cycles of nutrients in the Amazonian forest, we enter the realm of modeling. How difficult it will be to build a concrete exact and quantitative model of all the subcycles of nutrients in an (not to speak of 'the') Amazonian forest can be imagined when thinking of the tremendous number of plant and animal species, of different sizes and numbers of individuals, of different feeding specializations and of uneven distribution also within small areas of the hylaea, and when thinking of Frankens results that the chemistries of dropping waters and of stem-runoffs differ from species to species of trees ...

But in spite it can be useful to try to build a model of the cycling and sub-cycling on one or some selected and very limited places for to get at least some example of how the 'principle' of the closed nutrient circulation indeed works in reality; or to build another model without going into details, for to get an idea of the quantitative aspect of the nutrient circulation in the forest subdivided not into units of circulator species of plants and animals but subdivided perhaps into 'compartments' (stories) (cf. Klinge 1973b).

In both cases, however, basing oneself on one or the other of both models, it will not be possible to make exact predictions about concrete and quantitative consequences of interferences in that ecosystem: in the first case, because the cycling and sub-cycling in a selected small area is never representative for a greater one of that extremely diverse system since every small area may be different from the next one and since the studied small area is also 'open' to all sides and actively and passively connected with the unstudied and somehow different surrounding; in the second case, because a compartment is a 'black box' of which we do not understand the mechanism inside.

In both cases we can only say that it makes a difference whether an interference interrupts only a small nutrient cycle or homoeostatic circle through which only an insignificant fraction of the total metabolized amount of matter circulates or of energy flows, e.g. when only one small and unfrequent animal species be extinct − that would correspond to cutting or ligaturing a capillary −, or whether a main artery be opened through which the greatest proportion of circulating matter and of energy flows and is being distributed; that would happen when e.g. the most important and numerous trees of the forest be felled and exported or burnt.

Besides of their role as 'capillaries' or as greater vessels for the circulation of matter and the flux of energy through an ecosystem, however, small and unfrequent species also may exert a 'trigger function' for the survival or death of 'main arteries', e.g. by pollinating the flowers of a big and frequent tree species which without them would no more be able to propagate itself, resp., for taking another and imagined example, by transmitting an important disease to a pollinator of such an important tree species, etc. etc.

But also without going into details of the closed circulation principle, its splitting into sub-cycles of different sizes etc., that system in itself implicates the impossibility of lasting harvesting from it, for exportation, any appreciable amounts of organic products. Harvesting would mean the reduction of the quantity of circulating nutrients, and thus a ruinous exploitation. That would lead to a sooner or later break-down of the forest-ecosystem, corresponding to the constantly repeated removal of greater or smaller quantities of circulating nutrients which are not renewable. And here we come back to what has been said about the cutting of greater arteries or capillaries.

Felling and burning the forest, and equally harvesting for export, we see lead to the same end-effect the direction of which process, the substitution of an Amazonian forest-ecosystem by an impoverished one, is predictable when considering the three described original principles: poorness of nutrients, closed circulation of nutrients, and diversity.

Looking for examples which might represent such impoverished ecosystem in Amazonia we find the 'campos' on terra firme near Santarém which have probably been amplified and extended by former human activity (cf. Soili & Klinge, 1966) or the landscape of the Zona Bragantina East of Belém-Pará (Camargo l.c.; Egler 1961; Sioli 1973b).

Substituting the Amazonian high rainforest by a more campo- or steppe-like ecosystem will have its effect on still another 'principle' for the forest-ecosystem we have not yet mentioned. Bluntschli (l.c., p. 51) continues his characterization of the Amazonian landscape by saying: 'Der Kreislauf des Wassers vom Meere durch die Lüfte auf die waldige Erde und vom Wald durch die Stromebene wieder zum ewigen Meer, das ist das große, das Bild Amazoniens, sein Leben und sein Wesen beherrschende Moment. Vielleicht tritt nirgends sonst auf Erden mit solcher Klarheit und Sinnfälligkeit die mächtige Gewalt von Kreislauf der Wasser vor das geistige Auge des Menschen. In Amazonien gibt es nichts Totes und nichts Lebendiges, was nicht von ihm zeugen könnte'. ('The circulation of the water from the sea through the air to the wooded land, and from the forest through the plain of the huge river again to the eternal ocean, that is the great momentum which dominates the image of Amazonia, its life and its essence. Perhaps it is nowhere else on earth that the mighty force of the cycle of the waters appears before the spiritual eye of man with such clarity and evidence. In Amazonia there is nothing dead nor alive which could not testify for it.') (Translation by the author.)

But there is not only one big cycle of water in Amazonia, that carried from the sea with the **tradewind** blowing over the **plain** (for to cite also the other ones of the four basic ecological characteristics outlined by Bluntschli) to the forest and back to the sea. We now know that it is only a smaller part, 44%, of the rains falling in Amazonia which is being brought in, as new water, by the tradewinds, while the greater rest, 56%, consists of recycled water (Molion 1976). And we know, too, that the evapotranspiration of a latifoliate forest is much higher than that of a non-forest vegetation, and that the surface runoff of rainwater is reduced to a minimum under dense forest cover compared with that under non-forest vegetation where it may reach near 100% (Kovda 1973).

Repeated cycling of great part of the available water is thus the fourth principle we discover when studying the vast Amazonian forest ecosystem. And with it it becomes easy to predict the direction of changes which will happen in Amazonia when the forest will be cleared in large extensions.

The high rate of recycled water warrants the relative constant humidity of the climate during the course of the year, even when almost all of Central Amazonia belongs to the *Am* climatic type in the Köppen classification. I.e. that the climate is near monsoon character, with a stronger rainy season whereas the dry season is never long nor pronounced. Only where a zone of a bit drier climate, of *Aw* type, advances from the Northwest towards the middle lower Amazon around Santarém, there the annual dry season is more remarkable, with now and then in the course of the years up to 4 weeks without rain. And just there we see those 'campos' I mentioned above.

Applying this fourth principle of the Amazonian forest ecosystem, the recycling of great part of the rain water achieved by the peculiarity of the forest vegetation, to possible effects of interferences, we can conclude that an extensive deforestation will, by reducing the recycled portion of the rainwater through reduced evapotranspiration and increased surface runoff, not only diminish the total annual amount of precipitation but simultaneously intensify the seasonal periodicity of the pluvial climate, i.e. assimilate it to that of cerrado-covered Central Brazil. And while the groundwater level in Amazonia is generally deep so that the roots of the forest trees often cannot reach it and therefore depend on the continuous supply of rain, the question arises as to whether young forest trees will be able to take over again abandoned deforested (and impoverished) areas after a climatic change has happened, and even whether the remaining areas of original forests can stand that climatic alteration. That capacity of remaining forests, as well as that of young forests, will differ from area to area and from tree species to tree species since it will not depend only on the changed pluviosity but on many other factors, too, as there are groundwater level, water storage capacity of the soils, temperature of the denuded soil surface, physiological properties of the tree species, etc. etc. Here, an answer, but for a locally very restricted area only, could perhaps be given by a complicate model based on very many studies on very many items of just that restricted uniform area and of the expected tree species.

Would this be worthwhile? Experience teaches that, when decision-makers who come to their decisions first of all by emotional reasons (desire for business, or for power, what is the same), really ask scientists for expectable consequences of some planned interference in nature by, e.g., a 'development' project, the scientists generally need many years for to produce a correct answer. In the meantime that project has since long been carried out ...

I hope to have demonstrated that establishing 'principles' and building 'models' are two ways in the endeavour to understand the functioning of ecosystems. Both have their advantages and disadvantages and should be used according to our aims and possibilities when we try to analyse an ecosystem which is an other thing than just the sum of its parts and also doesn't consist only in more of less complicate matter and energy pathways as they are sometimes presented in impressive circuit-schemes. An ecosystem also comprises the whole colourfulness of life and the overwhelming richness in fates which are borne by it and make its content. And just thereby an ecosystem becomes really interesting and exciting for a biologist. But these properties are beyond principles and models.

References

Benz, R. 1923. Die Stunde der deutschen Musik I. Jena. Cited from Benz, R. (1977): Dem Geiste ein Haus. Sigmaringen.

Bluntschli, H. 1921. Die Amazonasniederung als harmonischer Organismus. *Geogr. Z.* 27: 49-67.

Camargo, F.C. de 1949. Terra e colonização no antigo e novo quaternário. *Bol. Mus. Par. E. Goeldi* 10: 123-147.

Egler, E.G. 1961. A Zona Bragantina no Estado do Pará. *Rev. Bras. de Geogr.* 23: 527-555.

Falesi, I.C. 1972. O estado atual dos conhecimentos sôbre os solos da Amazônia Brasileira. In: Zoneamento Agricola da Amazônia – 1ª Aproximaçao. *Bol. Técn. IPEAN* 55: 1-67.

Fittkau, E.J. 1971. Ökologische Gliederung des Amazonasgebietes auf geochemischer Grundlage. *Münster. Forsch. Geol. Paläont.* 20/21: 35-50.

Fittkau, E.J. 1973. Artenmannigfaltigkeit amazonischer Lebensräume aus ökologischer Sicht. *Amazoniana* 4: 321-340.

Furch, K. 1976. Haupt- und Spurenmetallgehalte zentralamazonischer Gewässertypen. Biogeographica 7: 27-43.

Gerwin, R. 1978. Entstehen Eiszeiten per Zufall? *MPG Spiegel* 1: 1-3.

Hedgpeth, J.W. 1977. Models and Muddles. Some philosophical observations. *Helgoländer wiss. Meeresunters.* 30: 92-104.

Herrera, R., Merida, T., Stark, N. & Jordan, C.F. 1978. Direct Phosphorus Transfer from Leaf Litter to Roots. *Naturwissenschaften* 65: 208-209.

Jansson, B.-O. 1976. Modeling of Baltic Ecosystem. *Ambio Special Report* 4: 157-169.

Katzer, F. 1903. Grundzüge der Geologie des unteren Amazonasgebietes (des Staates Pará in Brasilien). Leipzig.

Klinge, H. 1973a. Root mass estimation in lowland tropical rain forests of Central Amazonia, Brazil. I. Fine root masses of a pale yellow latosol and a giant humus podsol. *Trop. Ecol.* 14: 29-38.

Klinge, H. 1973b. Struktur und Artenreichtum des zentralamazonischen Regenwaldes. *Amazoniana* 4: 283-292.

Klinge, H. & Fittkau, E.J. 1972. Filterfunktionen im Ökosystem des zentralamazonischen Regenwaldes, *Mitteilgn.. Dtsch. Bodenkundl. Gesellsch.* 16: 130-135.

Kovda, V. 1973. The Soils of the Earth Planet and Man's Activities. In: Sioli, H. (ed.): Ökologie und Lebensschutz in internationaler Sicht. Freiburg: 63-90.

Molion, L.C.B. 1976. Possiveis efeitos de um deflorestamento em grande escala no clima da Amazônia. Ciência e Cultura, Suplem., Resumos 28° Reun. an. Soc. Bras. Progr. Ciência: 406.

Schmidt, G. 1972. Chemical properties of some waters in the tropical rain-forest region of Central Amazonia. *Amazoniana* 3: 199-207.

Schwabe, G.H. & Klinge, H. 1960. Gewässer und Boden als Forschungsgegenstand. *An. Edafol. y Agrobiol.* 19: 519-568.

Sioli, H. 1950. Das Wasser im Amazonasgebiet. *Forsch. u. Fortschr.* 26: 274-280.

Sioli, H, 1954. Beiträge zur regionalen Limnologie des Amazonasgebietes. II. Der Rio Arapiuns. *Arch. f. Hydrobiol.* 49: 448-518.

Sioli, H. 1955. Beiträge zur regionalen Limnologie des Amazonasgebietes. III. Über einige Gewässer des oberen Rio Negro-Gebietes.*Arch. f. Hydrobiol.* 50: 1-32.

Sioli, H. 1963. Beiträge zur regionalen Limnologie des brasilianischen Amazonasgebietes. V. Die Karbonstreifen Unteramazoniens. *Arch. f. Hydrobiol.* 59: 311-350.

Sioli, H. 1972. Ökologische Aspekte der technisch-kommerziellen Zivilisation und ihrer Lebensform. Biogeographica 1: 1-13.

Sioli, H. 1973a. Introduction into the Problems: The Situation of Modern Civilization in the Light of the Ecological Aspect of Life. In: Sioli, H. (ed.): Ökologie und Lebensschutz in internationaler Sicht. Freiburg: 9-34.

Sioli, H. 1973b. Recent Human Activities in the Brazilian Amazon Region and their Ecological Effects. In: Meggers, B., Ayensu, E. & Duckworth, W.D. (eds.): Tropical Forest Ecosystems in Africa and South America: A Comparative Revue. Washington: 321-334.

Sioli, H. 1977. Amazonasgebiet – Zerstörung des ökologischen Gleichgewichtes? *Geol. Rundschau* 66: 782-795.

Sioli, H. & Klinge, H. 1966. Anthropogene Vegetation im brasilianischen Amazonasgebiet. In: Tüxen, R. (ed.): Anthropogene Vegetation. Den Haag: 357-367.

Sombroek, W.G. 1966. Amazon Soils. Wageningen.

Stark, N. 1969. Direct nutrient cycling in the Amazon Basin. 11° Simpósio y Foro de Biologia Tropical Amazónica: 172-177.

Author's address:
Prof. Dr. Harald Sioli
Max-Planck-Institut für Limnologie
Abteilung Tropenökologie
Postfach 165
2320 Plön, FRG

VERBREITUNGSTYPEN BRASILIANISCHER BLATTSCHNEIDER-AMEISEN DER GATTUNG ATTA

H. Troppmair & P. Müller

Wegen ihrer landwirtschaftlichen Bedeutung bereits seit dem 16. Jahrhundert bekannt und trotz Entwicklungen zahlreicher Bekämpfungsmethoden auch in den Kulturlandschaften immer noch vorhanden, rücken die neotropischen Blattschneiderameisen der Gattung Atta zunehmend auch in den Mittelpunkt autökologischer und ökosystemarer Analysen (Amante 1973, Goncalves 1960, Mariconi 1970, Troppmair 1973). Von den bekannten 16 neotropischen Atta-Arten kommen 9 auch in Brasilien vor. Ihre Areale folgen in verblüffender Weise den Arealzentren, die von zahlreichen Autoren in jüngster Zeit für andere Invertebraten (u.a. Brown et al. 1974, Spassky et al. 1971, Turner 1972, Winge 1973), Vertebraten (u.a. Haffer 1969, Müller 1970, 1973, 1977, Müller & Schmithüsen 1970, Vanzolini und Williams 1970, Vuilleumier 1975) aber auch Pflanzenarten (u.a. Prance 1973) beschrieben wurden.

Obwohl die in Brasilien vorkommenden Arten taxonomisch gut definierbar sind (*A. sexdens*, *A. laevigata*, *A. bisphaerica*, *A. opaciceps*, *A. cephalotes*, *A. capiguara*, *A. goiana*, *A. robusta*, *A. vollenweideri*), wissen wir trotz zahlreicher autökologischer Untersuchungen noch wenig über ihre Ausbreitungsdynamik, die Lage ihrer Ausbreitungszentren und ihre phylogenetischen Verwandtschaftsbeziehungen. Eine vergleichende Betrachtung ihrer Arealsysteme läßt jedoch vermuten, daß ihre Arealgeschichte eng gebunden ist an die Geschichte der südamerikanischen Landschaften. Als Primärkonsumenten sind sie an Pflanzenformationen gebunden. Exemplarisch kann das für die Arten *A. laevigata* (Campo Cerrado) und *A. cephalotes* (Regenwälder) gezeigt werden.

Atta laevigata kommt als Art der offenen Landschaften nicht nur in den Campos Cerrados Zentralbrasiliens, sondern auch auf den isolierten Campo-Inseln der amazonischen Hylaea und in den Llanos des Orinokos vor. Temperatur und Niederschläge bestimmen ebenso ihre Tages- und Jahreszeiten-Periodik wie die Phänologie der für ihre Pilzkulturen notwendigen Dikotyledonen. Bei Temperaturen oberhalb 18° ist die Art besonders aktiv (u.a. Blattschnitt und Erdauswurf). Unterhalb von 18° (bis 13°) kommt es zu einer bis zu 50%igen Reduktion der Arbeitstätigkeit. Während der winterlichen Trockenzeit sind die Tiere auch tagaktiv, während der wärmeren Regenzeit überwiegend nachtaktiv. Bis zu 350 kg Blattmaterial kann von einem Zweimillionen-Volk pro Jahr eingetragen werden. Die Nester werden auf tiefgründigen, gut durchlüftet Latosolen (pH 4-5) bis in 10 Meter Tiefe gefunden. Die Anlage der Pilzkulturen folgt dabei einem Temperatur- und Feuchtigkeitsgradienten. Drei Jahre nach der Nestgründung durch eine

Königin erfolgt der erste Hochzeitsflug frischgeschlüpfter Königinnen (Troppmair 1973) in max. 1,5 bis 3 km Entfernung vom alten Nest. Die Nestdichte schwankt naturgemäß erheblich. Auf agrarisch ungenutzten Flächen Zentralbrasiliens beträgt die durchschnittliche Nestdistanz 20 Meter.

Atta cephalotes ist im Gegensatz zu *A. laevigata* streng an die Regenwälder gebunden. Ähnlich wie andere Waldarten (z.B. *Lachesis mutus, Bothrops bilineatus*) zeichnet sich ihr disjunktes Areal durch ein Teilareal in Amazonien und der Serra do Mar aus. Die Art ist wesentlich lichtempfindlicher als *A. laevigata* und überwiegend nachtaktiv. Beide Arten sind, obwohl völlig andere Nischen als z.B. Vogel- oder Reptilienarten einnehmend, in ihrer Verbreitung weitgehend homotop zu den Makro-Ökosystemen 'Hylaea' und 'Campo Cerrado'. Deshalb muß ihre Arealgenese mit der Landschaftsgeschichte der tropischen Regenwälder bzw. Savannen korrelierbar sein.

Wie weit die voneinander isolierten Populationen unterschiedlich differenziert sind, ist bisher noch ungeklärt, da elektrophoretische Untersuchungen zur Aufklärung ihres Allelpolymorphismus noch ausstehen (vgl. Müller & Steiniger 1978).

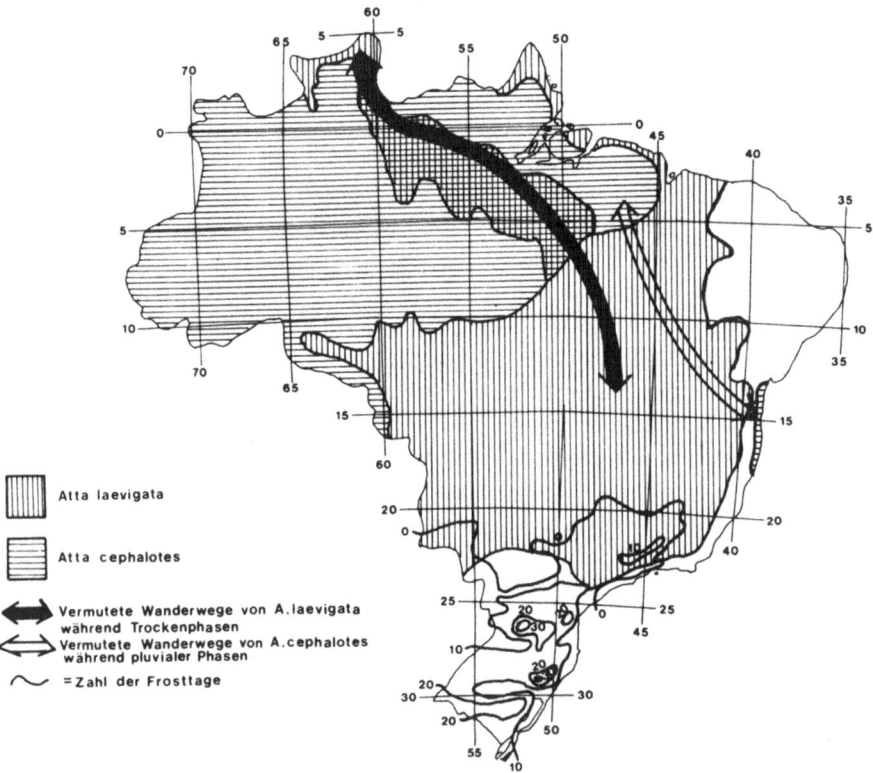

Abb. 1. Verbreitung von *Atta laevigata* und *Atta cephalotes* in Brasilien und ihre vermuteten Wanderwege.

Dennoch zweifeln wir nicht daran, daß die Arealdisjunktionen der beiden Arten durch postglaziale Vegetationsfluktuationen interpretiert werden können, wie sie analog für andere neotropische Organismen bereits nachgewiesen wurden.

Literatur

Amante, E. 1973. Influência de alguns fatores microclimaticos sobre a formiga sauva Atta laevigata, Smith 1858, Atta sexdens rubripilosa, Forel 1908, Atta capiguara, Goncalves 1944 (Hymenoptera, Formicidae) em formigueiros localizados no Estado de São Paulo. São Paulo.

Brown, K.S., Sheppard, P. & Turner, J. 1974. Quaternary refugia in tropical America: evidence from race formation in Heliconius butterflies. *Proc. R. Soc.* 187: 369-378, London.

Goncalves, A. 1960. Distribuição, biologia e ecologia das sauvas. In: Divulgação Agronômica 1.

Haffer, J. 1969. Speciation in Amazonian Forest Birds. *Science* 165: 131-137.

Mariconi, F. 1970. As saúvas. Agronomica Ceres. São Paulo.

Müller, P. 1970. Vertebratenfaunen brasilianischer Inseln als Indikatoren für glaziale und postglaziale Vegetationsfluktuationen. *Abhdl. Dtsch. Zool. Ges. Würzburg* 1969: 97-107.

Müller, P. 1973. The Dispersal Centres of Terrestrial Vertebrates in the Neotropical Realm. Biogeographica 2, Junk, The Hague.

Müller, P. 1977. Tiergeographie. Teubner, Stuttgart.

Müller, P. & Schmithüsen, J. 1970. Probleme der Genese südamerikanischer Biota. Festschr. Gentz, Hirt, Kiel.

Müller, P. & Steiniger, H. 1978. Evolutionsgeschwindigkeit, Verbreitung und Verwandtschaft brasilianischer Erdleguane der Gattung Liolaemus (Sauria, Iguanidae). Mitt. Schwerpunkt Biogeographie 9, Univ. Saarbrücken.

Prance, G.T. 1973. Phytogeographic support to the theory of Pleistocene forest refuges in the Amazon Basin, based on evidence from distribution patterns in Carycaraceae, Chrysobalamaceae, Dichapetalaceae and Lecythidaceae. *Acta Amazonica* 3 (3): 5-28.

Spassky, B. et al. 1971. Geography of the Sibling Species related to *Drosophila willistoni*, and of the Semispecies of the *Drosophila paulistorum* complex. *Evolution* 25 (1): 129-140.

Troppmair, H. 1973. Estudo zoogeográfico e ecológico das formigas do Genero Atta (Hymenoptera) com ênfase sobre a Atta laevigata, Smith 1858, no Estado de São Paulo. Unesp, Rio Claro.

Turner, J. 1972. The genetics of some polymorphic forms of the butterflies *Heliconius melpomene* (Linnaeus) and *H. erato* (Linnaeus). II. The hybridiziation of subspecies *H. melpomene* from Surinam and Trinidad. *Zoologica* 56: 125-157, New York.

Vanzolini, P.E. & Williams, E.E. 1970. South American Anoles: The geographic differentiation and evolution of the *Anolis chrysolepis* species group (Sauria, Iguanidae). *Arq. Zool.* 19: 1-298.

Vuilleumier, F. 1975. Zoogeography. In: Avian biology 5: 421-496. Acad. Press, New York.

Winge, H. 1973. Races of Drosophila willistoni sibling species: Probable origin in quaternary forest refuges of South America. *Genetics* 74: 297-298.

Address of the authors:

Prof. Dr. H. Troppmair
Campus de Rio Claro
Estado de São Paulo
Caixa Postal 178

Prof. Dr. P. Müller, Lehrstuhl für Biogeographie
Geographisches Institut
Universität des Saarlandes
66 Saarbrücken, FRG

PHYSISCH-ANTHROPOGEOGRAPHISCHE UNTERSUCHUNGEN AN DER OSTAFRIKANISCHEN POPULATION DER ELMOLO AM LAKE TURKANA – EIN BEITRAG ZUR BIOGEOGRAPHISCHEN FORSCHUNG

G. Kenntner & E. Ludwig

Abstract

The Elmolo Population at the Lake Turkana is in a physical state which is extremely bad. Deformations at the skeleton are to be found very often. Besides the behaviour of the specific people most of all are to be blamed the ecological conditions of their living space, poor nourishment and insufficient hygienic conditions.

The main cause certainly lies in the one-sided and deficient food of the Elmolo. An analysis of fish and water has proved that a bad constellation of minerals is predominant in the food of the Elmolo which finds its expression especially in a lack of calcium. As against that you can determine high concentrations of fluorine and amounia. The suspicion upon fluorosis however cannot be confirmed. In all probability the deformations at the skeleton are a matter of lack of certain minerals causing rickets at an early or an old age. Besides the lack of minerals the bad hygienic conditions as well as certain habits (squating for hours, carrying heavy loads at the side of the body) are taken as responsible factors.

Because of the small number of individuals and the increasing migration to Loiyangalani a population that is anthropologically and medically very interesting, and a singular culture that is still nearly untouched and hardly known, will be gone before long. As well on humanitarian as on scientific reasons it is necessary to find and obtain by hard working ways and means to maintain this small population.

Der Mensch ist nicht nur die gestaltende Kraft seines Lebensraumes, die ihren Ausdruck in der Kulturlandschaft findet, er erfährt seinerseits auch Gestaltung durch die biotischen und abiotischen Faktoren des Raumes, in den er integriert ist.

So einfach sich dieser Aktions-Reaktions-Mechanismus im Bezugsfeld Mensch-Raum auch anhören mag, so kompliziert ist der Problemkreis, der sich dahinter verbirgt. Die geistigen und sozialen Charakteristika des Menschen spielen in diesem Zusammenhang ebenso eine Rolle wie der physisch-biologische Aspekt, wenn auch letzterer vielleicht naturwissenschaftlich exakter faßbar ist.

Kleine ethnische Gruppen in relativ naturnahem Zustand in Isolationsräumen vermögen wohl am ehesten dazu beizutragen, Mensch-Raum-Beziehungen aufzuhellen, vor allem wenn sich der Einfluß der Umwelt so eigenartig bemerkbar macht wie im Falle der Elmolo vom Lake Turkana[1] in Nordkenia.

Der zahlenmässig 'kleinste Stamm Afrikas', wie die Elmolo vielfach genannt werden (McDougall 1974, Vagnby/Jacobs 1974, Spencer 1973), lebte bis in die

---

1 Seit 1976 offizielle Bezeichnung für den 1888 von Teleki und Höhnel entdeckten Rudolfsee.

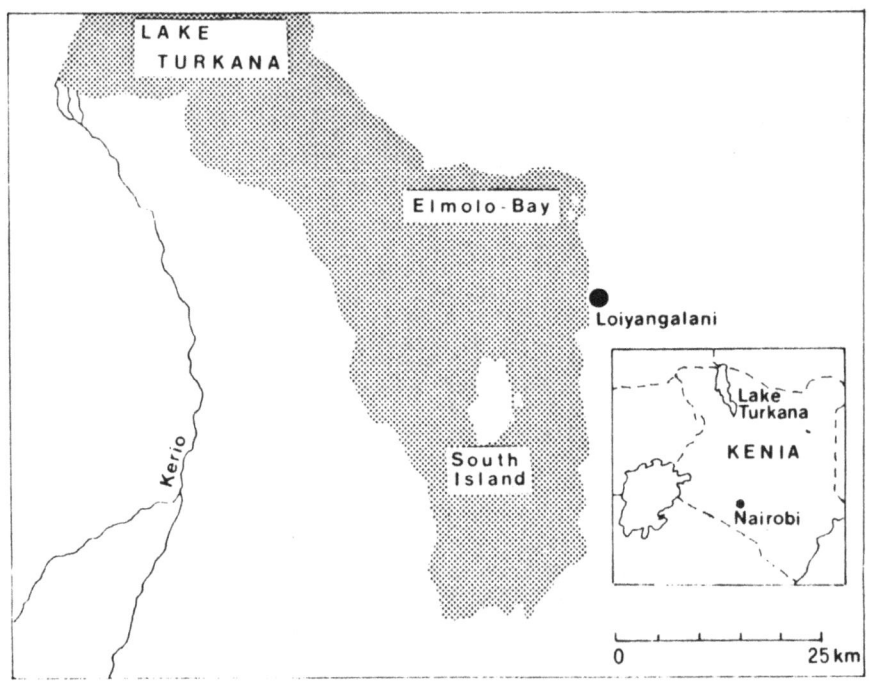

Abb. 1. Skizze Lake Turkana – Südteil.

jüngste Vergangenheit geographisch und kulturell vollkommen abgesondert von den nomadisierenden Nachbarstämmen der Rendille, Samburu und Turkana auf Inseln im südöstlichen Bereich des Lake Turkana (Elmolo-Bay) (Abb. 1, 2). Wenn auch gegenwärtig unter dem Einfluß der Missionsstation von Loiyangalani ein Teil der Elmolo ans Festland (Anderi) gezogen ist, wo in zunehmendem Maße Absorbierungsprozesse physischer (Mischehen) und kultureller (Sprache, Kleidung, Sozialstruktur) Art stattfinden, so besteht die autochthone Fischerpopulation auf den Inseln Lorian bzw. Koran weiter und bewahrt ihre Kultur und anthropologische Identität.

Die beiden Elmolo-Inseln werden je nach Wasserstand abwechselnd besiedelt. In Regressionsperioden bildet Lorian eine Landzunge aus, so daß es ohne Schwierigkeit möglich ist, zu Fuß vom Festland auf die Insel zu gelangen. In diesem Falle verlegen die Elmolo ihre Siedlung auf Koran, um vor zudringlichen Fremden oder Feinden geschützt zu sein. Umgekehrt kehren sie bei Wasserspiegeltransgression auf Lorian zurück. Nach unseren Beobachtungen lebten die Elmolo 1972 auf Koran, 1975 auf Lorian und 1977 wieder auf Koran (Abb. 3, 4).

Aufgrund somatologischer Untersuchungen läßt sich der Elmolo-Typus wie folgt charakterisieren: Die Elmolo sind mit einer durchschnittlichen Körperhöhe von 170,8 cm bei den Männern und 163,9 cm bei den Frauen mittelgroß bis groß. Auffallend ist der leptosome Körperbau und eine relative Langbeinigkeit (ske-

BEWOHNTE SIEDLUNG
EHEMALIGE SIEDLUNG

LORIAN
KORAN
ELMOLO
BAY
ELMOLO CAMP

ANDERI LOIYANGALANI

SOUTH
ISLAND

N

0   10   20 km

Abb. 2. Skizze Siedlungen am südöstlichen Lake Turkana.

165

Abb. 3. Nur durch einen schmalen Wasserstreifen ist Lorian derzeit vom Festland getrennt. Am Beginn der Landzunge lag 1975 das Dorf der Elmolo.

Abb. 4. Von Lorian aus ist die Insel Koran, auf der sich heute das Elmolo-Dorf (1977) befindet, nur mit Flössen zu erreichen.

lischer Index[1] – Männer: 49,4; Frauen: 50,2). Das durchschnittliche Körpergewicht ist im Verhältnis zur Körperhöhe sowohl bei Männern (57,5 kg) als auch
bei Frauen (53,4 kg) niedrig. Der überwiegend mesokephale und nach hinten aus-
ladende Kopf besitzt ein schmales, langes Gesicht mit ausgeprägten negriden
Kennzeichen (Prognathie, wulstige Lippen), aber auch europiden Merkmalen. So
ist die Nase mittellang und mittelbreit, der Nasenrücken variiert von konvex bis
konkav. Die Hautfarbe ist dunkelbraun und das schwarze Haar mehr oder weniger
kraus und kurz (Abb. 5).

Die Untersuchung des Hautleistensystems der Elmolo ergab bei den Fingerbeer-
mustern der beiden Geschlechter ein Überwiegen von Schleifen (64,8 %) gegen-
über Wirbeln (29,0 %) und Bögen (6,2 %).

Serologische Untersuchungen ergaben im ABO-System die Phänotypen 0 mit
49% und B mit 40%. Die Blutgruppen A und AB waren dagegen anteilmäßig nur
gering vorhanden. Diese Beobachtung wurde sowohl bei Männern als auch bei
Frauen gemacht.

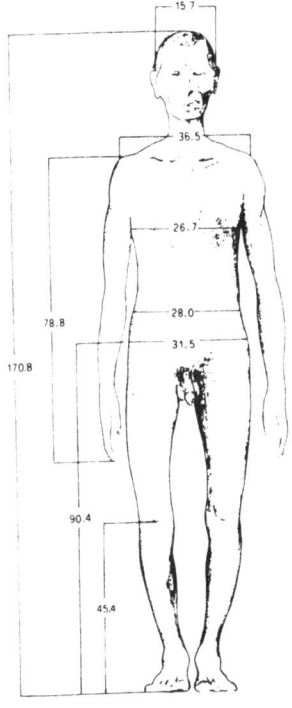
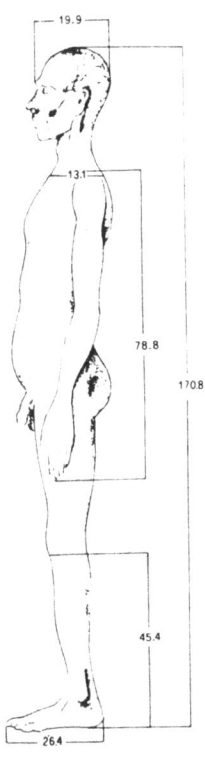

Abb. 5. Skizze Elmolo-Typus.

1 Skelischer Index = Sitzhöhe in % der Körperhöhe.

Serologische Merkmale sind ebenso wie die Merkmale des Hautleistensystems genetisch fixiert und in ihrer phänotypischen Ausprägung von keinerlei peristatischen Faktoren abhängig. Im Gegensatz zu den genetisch weitaus komplizierteren morphologischen Merkmalen und Merkmalskomplexen eignen sie sich in besonderem Maße für Gen-Analysen von Bevölkerungen (Schwidetzky 1962). Unsere anthropologischen Untersuchungen an den Elmolo werden daher vor allem bei rassengenetischen Fragestellungen von Bedeutung sein.

Die rassengenetische Beurteilung der Elmolo erweist sich als außerordentlich problematisch. Die Schwierigkeit liegt z.T. in dem Terminus 'Elmolo' begründet, der ursprünglich weder ein rassenkundlicher Begriff noch ein Stammesname war, wenn er auch heute als solcher verstanden wird. Das kuschitische Wort aus der Rendillesprache bezeichnet im Gegensatz zu den viehbesitzenden Nomaden die besitzlosen Bevölkerungsteile am Lake Turkana, die 'Armen' oder 'Verarmten', die gezwungen sind, vom Fischfang zu leben (Spencer 1973, Vagnby/Jacobs 1974, Hillaby 1964, Willock 1974).

Die Bezeichnung von Fischerpopulationen am Lake Turkana als 'Elmolo' allein aufgrund ihres Wirtschaftsstatus hat der rassengenetischen Beurteilung der heutigen Elmolo 2 Alternativen eröffnet:

1. Die Elmolo sind eine Mischpopulation aus verarmten Nomaden unterschiedlicher Herkunft

Abb. 6. Schwerwiegende Knochendeformierungen an Beinen (Säbelscheidentibia) und Thorax sind bei Frauen in auffälliger Häufigkeit zu beobachten.

(Samburu, Rendille, Shangilla) ohne anthropologische Eigenständigkeit.

2. Die Elmolo sind Überreste einer Urbevölkung, die durch später eingewanderte Nomadenvölker in ihren Lebensraum (Inseln) abgedrängt wurden.

So klar diese Alternativen – Elmolo als Nachfahren verarmter Viehzüchter oder Elmolo als Reste einer Urbevölkerung – in der Theorie auch sein mögen, in der Praxis wird eine exakte rassengenetische Beurteilung durch das Phänomen der Rassenmischung einerseits und der Isolation andererseits zusätzlich erschwert. Bestehen bleibt ein noch weitgehend ungelöstes Problem von hohem wissenschaftlichen Interesse, das umfangreiche vergleichende Untersuchungen an Skelettserien und lebenden Elmolo sowie ihren Nachbarvölkern erfordert.

Abb. 7. Auch bei jungen Männern und Kindern sind bereits 'Bumerang'-Beine zu beobachten. (Diese Beine gehören einem 25 Jahre alten Mann).

169

So interessant die Frage der rassengenetischen Einordnung der Elmolo auch sein mag, ein anderes Problem scheint vorrangig: Wer die Elmolo auf ihrer Insel besucht, wird die erschreckende Feststellung machen, daß sich die Population in einem auffallend schlechteren physischen Zustand befindet. Allgemein sind Anzeichen schwerwiegender Mangelernährung zu erkennen. Aufgrund unserer Beobachtungen und Untersuchungen läßt sich der allgemeine physische Zustand folgendermaßen darstellen:

Kinder und Erwachsene beiderlei Geschlechts weisen häufig Skelettdeformationen auf. Schon bei älteren Kindern ist in Kombination mit den sogenannten O-Beinen die Säbelscheidentibia feststellbar. Dabei handelt es sich um eine Verbiegung der Schienbeine nicht wie üblicherweise nach hinten, sondern nach vorne ('Bumerang'-Beine). Hinzu kommen mit zunehmendem Alter Veränderungen an den Kniegelenken, die durch Muskelschwund im Wadenbereich noch deutlicher in Erscheinung treten (Abb. 6, 7).

Abb. 8. Während bei Kindern und jungen Leuten Körperhaltung und Gang noch normal sind, haben ältere Leute häufig eine nach vorn geneigte Körperhaltung, verbunden mit rachitischen Veränderungen am Thorax.

170

Häufig sind bei Erwachsenen rachitische Veränderungen am Thorax zu beobachten (Harrinson'sche Furche). Hierbei handelt es sich um eine Abflachung und horizontale Einbuchtung der seitlichen Thoraxpartien. Außerdem sind bereits bei jungen Leuten Veränderungen an den Handgelenken festzustellen. Erwachsene beiderlei Geschlechts haben ungewöhnlich häufig einen Buckel mit gleichzeitiger seitlicher Verkrümmung der Wirbelsäule (Kyphoskoliose). Der nach vorne geneigte Körper und der Watschelgang deuten darauf hin, daß auch die Hüftgelenke geschädigt sind (Abb. 8). Manche Personen machten durch deutlich in Erscheinung tretende Knochendeformierungen und eine beidseitige Erblindung einen äußerst bemitleidenswerten Eindruck. Einige ältere Menschen konnten sich aufgrund ihrer

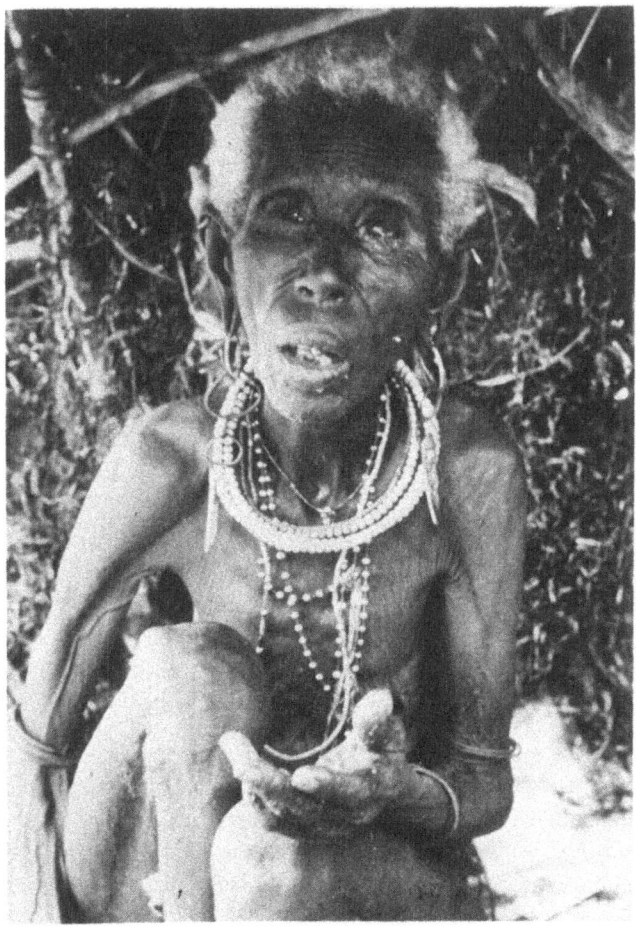

Abb. 9. Die vier auf der Insel noch lebenden alten Frauen sind allesamt erblindet. Ausgemergelt und weitgehend bewegungsunfähig sitzen sie in Einzelhütten und sich volkommen auf die Hilfe ihrer Stammesgenossen angewiesen.

schwerwiegenden Skelettdeformierungen nur noch in hockender Stellung vorwärtsbewegen oder waren fast vollkommen bewegungsunfähig (Abb. 9).

Diese Phänomene sind so auffällig, daß sie auch früheren Besuchern der Elmolo nicht entgangen sind und immer wieder Erwähnung fanden: Bereits Hoey (1911) bemerkt 'the presence of some deformities in most of them'. Bis in die Gegenwart hinein wird in der Fachliteratur laufend auf den schlechten physischen Zustand der Elmolo, insbesondere auf Knochendeformierungen, hingewiesen (Lloyd-Jones 1925, Worthington 1932, Henderson 1958, Adamson 1960, 1967, 1969, Graham/Beard 1973 u.a.).

Die wohl eindringlichsten Beschreibungen liefern Dyson/Fuchs (1937) und Spencer (1973):

These people suffer from an almost universal deformity, a forward and sometimes outward bowing of the shin bones, which may reach such a degree that makes walking a difficulty and gives the leg the appearance of having a second knee somewhere above the ankle. It seems almost certain that the deformity is due to a diet deficiency. They live entirely on fish, crocodiles, and turtles that are caught in the lake, and the only water they drink is of the lake. It is known that there is little or no calcium in the lake water, as it is precipitated by the soda-rich waters. It therefore seems possible that not only the water but the animals living in it may suffer from a calcium deficiency, and in this case humans dependent upon the lake for their food and water would probably lack a sufficient supply of calcium (Dyson/Fuchs 1937).

To any visitor, the most noticeable feature of the Elmolo is their diseased appearance: their lips are blotchy and their teeth discoloured; apart from those born with deformities, younger men complain of weak legs, middle-aged men are distinctly bow-legged, an older people can no longer walk (Spencer 1973).

Anm.: Darüber hinaus treten noch weitere Krankheiten auf. So waren bei zwei Männern auf der Insel starke Lymphstauungen zu beobachten (Elephantiasis arabum). Hierbei handelt es sich um eine Verstopfung der Lymphbahnen durch Filariden. Auch Hautkrankheiten und Zahnanomalien sind nicht selten anzutreffen.

Obwohl die Notwendigkeit von Kausalitätsforschungen bereits sehr früh erkannt und auch postuliert worden ist (Dyson/Fuchs 1937) und manche Theorie in diesem Zusammenhang geäußert wurde (Spencer 1973), ist bisher keine wissenschaftliche Untersuchung erfolgt, um mögliche Ursachen und Wirkungsprinzipien aufzudecken.

Die symmetrische Verteilung der Skelettverformung auf beide Körperseiten sowie ihr häufiges Auftreten in der Population (ca. 30%) deuten darauf hin, daß es sich hier um eine Erkrankung des Knochensystems handelt, die offensichtlich mit den Lebensgewohnheiten und den extremen Lebensbedingungen der Elmolo in Zusammenhang gebracht werden muß. Die Umwelt der Elmolo mit allen direkt und indirekt auf den Menschen einwirkenden Faktoren rückt damit in den Mittelpunkt der Betrachtung:

Der Lake Turkana liegt in einer trockenen, wüstenhaften Vulkanlandschaft im nördlichen Teil des ostafrikanischen Grabens in 370 m Höhe NN. Die südwestliche Uferregion ist eine der lebensfeindlichsten Gegenden überhaupt, ein Gebiet von extremer Trockenheit mit Niederschlagsmengen unter 200 mm/Jahr und Temperaturen von über 40°C im Schatten. Der fast vegetationslose Boden ist von dunklen Vulkanbrocken übersät, die der Landschaft ein äußerst trostloses Aussehen verleihen. Die scharfkantige Lava reicht bis an das brackige Seewasser heran, das von dem tagsüber ständig wehenden SO-Wind aufgepeitscht wird (Abb. 10). Dieser Wind kann bisweilen Sturmstärke erreichen und ein Befahren des Sees zu einem lebensgefährlichen Unternehmen machen (Fuchs 1935).

Einen erfreulichen Anblick in dieser Einöde bietet die Oase Loiyangalani (Abb. 1, 2) mit ihren Palmhainen. Sie wird vom Niederschlagswasser des 2700 m hohen Mt. Kulal versorgt, ein Bergmassiv im Osten von Loiyangalani.

Die Borassus-Palmen (Borassus aethiopum) Loiyangalanis liefern den Elmolo wichtige Materialien: Stämme für ihre Flösse, Palmwedel für den Hüttenbau, Gestänge und Fasern für Waffen und Geräte (Fischkörbe, Netze, Matten u.a.).

Im Gegensatz dazu macht die dunkle, vegetationslose Steinmasse der Elmolo-Inseln einen eher abschreckenden Eindruck. Die ca. 20-30 m über den See aufragenden Vulkaninseln von je 1/2 qkm Größe liegen etwa 1 km vom Festland entfernt in der Elmolo-Bay, 10 km nördlich von Loiyangalani (Abb. 2).

Das ausgetrocknete Gebiet um den See stellt die dort verstreut lebenden nomadisierenden Stämme (Rendille, Samburu, Turkana, Gabbra u.a.) ständig vor Existenzprobleme. Die seßhaften Elmolo dagegen haben keine quantitativen Ernährungssorgen; sie decken ihren Bedarf an Wasser und Nahrung nahezu vollkommen und ausreichend aus dem See. Fast ausschließliches Nahrungsmittel ist der Fisch, insbesondere Nilbarsch und Tilapien, gelegentlich ergänzt durch Krokodil- und Schildkrötenfleisch.

Der Fischfang wird von den Elmolo-Männern während der Nacht oder frühmorgens betrieben. Es werden im wesentlichen zwei Fangmethoden praktiziert: Nilbarsche und Krokodile erlegt man mit der Harpune, kleinere Fische werden mit dem Fischspeer oder mit dem Netz gefangen. Gelegentlich findet auch ein Fischkorb Verwendung, der im seichten Wasser über den Fisch gestülpt wird, bevor das Tier mit einem Fischspeer oder einem Antilopenhorn den Todesstoß erhält. Im tiefen Wasser wird vom Palmstammfloß, dem wichtigsten Verkehrs- und Transportmittel der Elmolo, aus gejagt. Dabei bilden zwei bis drei Elmolo gewöhnlich eine Jagdgemeinschaft, die bei dem außerordentlichen Fischreichtum des Sees stets erfolgreich ist und pro Jagdzug für einen Vorrat für mehrere Tage sorgt (Abb. 11, 12, 13).

So umfangreich die Ernährung der Elmolo auch sein mag, so wenig abwechslungsreich ist anderseits ihr täglicher Speiseplan, auch wenn neuerdings die Fischnahrung durch Maismehl (Ugalli) ergänzt wird, das in Loiyangalani käuflich zu erwerben ist (Abb. 12).

Das Ausnehmen und Vorbereiten der Fische für den Kochtopf ist Frauenarbeit, während beim Ausschlachten des Nilbarsches auch Männer zu beobachten waren (Abb. 14).

Nur das Filet wird verzehrt, der Rest wird weggeworfen und verfault zwischen den Hütten. Kleinere Fische werden entweder auf der Feuerstelle zwischen den heißen Steinen geröstet oder mit Wasser zu einer Fischsuppe verkocht. Das Nilbarschfleisch schneiden die Elmolo in Streifen und trocknen es an der Sonne. Gekocht wurde bis vor kurzem noch in selbstgefertigten Tontöpfen, gegessen wurde aus Schildkrötenpanzern. Der traditionelle Hausrat weicht jedoch zunehmend Blechbüchsen, die durch Fremde in die Hand der Elmolo gelangen (Abb. 18).

Der tägliche Bedarf an Flüssigkeit kann nur durch das Seewasser gedeckt werden (Abb. 14). Dieses ist für Nichteinheimische wegen seines hohen Sodagehaltes ungenießbar und zieht häufig Darmerkrankungen nach sich (Höhnel 1892).

An einem quantitativen Nahrungsmangel leiden die Elmolo keinesfalls; Fisch steht in unübersehbarer Fülle zur Verfügung. Es stellt sich jedoch die Frage, ob

173

Abb. 10. Alles würde man in dieser trostlosen Landschaft aus Lavabrocken erwarten, nur nicht einen See von diesem Ausmaß. Der Lake Turkana ist mit einer Fläche von 8500 qkm 16 mal so groß wie der Bodensee. Am Horizont im Westen zeichnen sich die dunklen Vulkankegel von South Island, der großen Südinsel, ab.

Abb. 11. Die Elmolo benutzen beim Jagen von Fischen eine Harpune.

Abb. 12. Die sehr einfach konstruierten Flösse der Elmolo bestehen aus 3-7 zusammengebundenen Stämmen der Borassus-Palme. Im seichten Wasser wird das Floß durch Staken vorwärts bewegt; im tiefen Wasser wird die Stange als Paddel benutzt. In den Säcken befindet sich Maismehl (Ugalli), das gerade nach Koran transportiert werden soll.

Abb. 13. In mondlosen Nächten waten 2 Elmolo-Jäger mit Speer und Harpune bewaffnet durch das seichte Wasser, um Krokodile zu jagen. Trotz des großen Krokodilreichtums ist die Jagd nicht immer erfolgreich und zudem eine gefährliche Angelegenheit, die viel Geschicklichkeit von den Jägern erfordert.

Abb. 14. Der Nilbarsch, der hier gerade filetiert wird, ist einer der beliebtesten Speisefische der Elmolo. Ihr Trinkwasser entnehmen sie einfach aus dem see.

und inwiefern ein qualitativer Nahrungsmangel besteht, der möglicherweise zur Erklärung des schlechten physischen Zustandes der Elmolo, insbesondere ihrer Knochendeformierungen beitragen könnte.

Infolge der ungünstigen Umweltbedingungen (Klima, Boden) fehlt pflanzliche Nahrung nahezu vollkommen, zumal die Elmolo nicht wie die Eskimo den vegetabilischen Mageninhalt ihrer Beutetiere verzehren. Vitaminhaltige Nahrung wie Milch, Milchprodukte und Eier sind ebenfalls nicht vorhanden.

So kommt es vermutlich zu einem Mangel an Vitaminen, vor allem an Vitamin D und an Kohlehydraten. Daß darüber hinaus auch Mineralstoffmangel besteht, belegen die nachfolgenden Analysen von Seewasser und Nilbarschtrockenfleisch.

Die chemische Analyse von getrocknetem Nilbarschfleisch (Tab. 1) ergibt beim Vergleich mit europäischem Speisefisch Unterschiede in den verschiedenen Substanzen, insbesondere in Bezug auf den Gehalt an Eisen, aber auch an Calcium und Phosphor: Der Mineralstoffgehalt beim Nilbarsch liegt insgesamt unter dem des Stockfisches. Der Unterschied im Wasser- und Fettgehalt ist auf den starken Austrocknungsvorgang beim Nilbarsch zurückzuführen. (Der Proteingehalt ist bei beiden Proben relativ hoch).

Die Wasseranalysen (Tab. 2) zeigen im Vergleich zu deutschen Proben ebenfalls einen geringeren Gehalt an Mineralstoffen, insbesondere an Calcium und Magnesium. Dagegen ist der Anteil an Natrium, Kalium sowie an Chlorid und Sulfat verhältnismäßig hoch. Auffallend ist auch die beträchtliche Konzentration von Fluor und Ammoniak. Der hohe Ammoniakgehalt kann darauf zurückgeführt

Tab. 1. Chemische Analyse von getrocknetem Nilbarsch, Lake Turkana (1977), im Vgl. zu Stockfisch (1977)

Substanz	Nilbarsch	Stockfisch
Wasser	11,6%	15,0%
Fett	1,9%	2,5%
Sonstige Mineralstoffe	4,9%	6,0%
Protein (N x 6,25)	82,9%	80,0%
Phosphor (P_2O_5)	1,8%	2,1%
Calcium (Ca)	0,04%	0,06%
Eisen (Fe)	0,002%	0,03%

Tab. 2. Analyse von Wasserproben entnommen vor der Ostküste der Elmolo-Insel Koran und vor der Küste von South Island 1977.

Substanz	Elmolo-Insel 15.7.1977 (Koran)	South Island 23.7.1977	Karlsruher Wasser 1.8.1977
$T_1 °C$	32,1	30,5	–
Ca^{2+} mg/l	5,7	–	124,0
Mg^{2+} mg/l	2,6	–	16,0
K^+ mg/l	23,15	–	3,6
Na^+ mg/l	950		11,2
F^- mg/l	0,89	0,87	(0,19)[*]
Fe^{3+} mg/l	0,05	0,18	0,02
NH_4^+ mg/l	3,34	0,03	–
NO_3^- mg/l	0,5	2,2	–
CL^- mg/l	1295,7	1287,5	50,4
PO_4^{3-} mg/l	4,0	3,8	–
SO_4^{2-} mg/l	394	370	–
$KH°$ dH	1,8	1,6	–
$GH°$ dH	59,4	67,2	–

werden, daß die Elmolo an derselben Stelle, an der sie ihr Trinkwasser entnehmen, auch ihre Fäkalien ins Wasser abgeben. Die starke Natriumkonzentration führt wahrscheinlich in Verbindung mit dem Sulfat (Na_2SO_4) zu chronischen Durchfallerscheinungen und damit zu beträchtlichen Nährstoffverlusten sowie zu einer Entwässerung des Körpers. Die Wasserverluste werden dann durch eine erhöhte Aufnahme von brackigem Seewasser (hoher Gehalt an Natriumchlorid) wieder regeneriert. Diese Vermutung findet dadurch Bestätigung, daß wir die Elmolo ungewöhnlich häufig beim Trinken beobachten konnten. Der relativ hohe Gehalt an Fluor legt den Verdacht nahe, daß die Elmolo unter einer Zahn- und Knochenfluorosis leiden. Hierfür könnten neben den verkrümmten Knochen auch ihre gelb-braun verfärbten Zähne sprechen, denn stärkere Fluorose führt zu einer Zerstörung des Zahnschmelzes und durch Einlagerung von Calciumfluorid zu Veränderungen von Knorpeln und Knochen. Nach Untersuchungen von Franke (1973) handelt es sich bei der 'Knochenfluorose' jedoch um eine 'Osteosklerose', die nicht wie bei den Elmolo mit Knochenverbiegungen, sondern mit Versteifungen im Wirbelsäulenbereich und in den großen Gelenken, wie dem Hüftgelenk, einhergeht. Dabei tritt nach Chlud (1973) erst dann eine Knochenschädigung ein, wenn im Trinkwasser mehr als 8 mg/l Na F auftritt und dieses Wasser über 25 Jahre hindurch ununterbrochen getrunken wird. Sehr schwere Knochenschäden sollen sogar erst bei Konzentration von 20 mg/l und mehr zu erwarten sein. Diese hohe Konzentration wurde jedoch von uns im Seewasser nicht festgestellt (Tab. 2). Auch genauere Zahnuntersuchungen zeigten, daß es sich bei den braunen Verfärbungen nur um dicke Zahnbeläge handelt, während der Zahnschmelz darunter nicht verändert ist.

Die Theorie, bei den Knochenverbiegungen der Elmolo handle es sich um eine Fluorosis (Spencer 1973), kann nach unseren Untersuchungen daher nicht bestätigt werden. Vielmehr muß angenommen werden, daß die ungünstige Mineralstoffkonstellation, vor allem der Mangel an Calcium und Vitamin D, den schlechten physischen Zustand der Elmolo bedingt.

Die Nahrung ist wohl derjenige Faktor, der den Menschen direkt und am nachhaltigsten beeinflußt und in dem eine Reihe von Faktoren der Natur- und Kulturumwelt zum Tragen kommen. Darüber hinaus kann der Mensch durch bestimmte Verhaltensweisen, die er sich in der Auseinandersetzung mit seiner Umwelt angeeignet hat, bestimmte Faktoren in ihrer Wirksamkeit unterstützen. Die hygienischen Verhältnisse, unter denen die Elmolo leben, müssen daher bei Erklärung der Deformierungserscheinungen ebenso berücksichtigt werden wie bestimmte alltägliche Lebensgewohnheiten.

Der Engländer Hillaby besuchte die Elmolo Anfang der 60er Jahre und fühlte sich in prähistorische Zeiten versetzt:

'Verfaulende Fischreste lagen auf der Erde herum, nackte Kleinkinder schlugen nach den Fliegen, an ein paar Stöckchen hingen zerlumpte Netze. Es war wie in der Jungsteinzeit. Mir kam es vor, als sei ich zufällig auf eine Rasse gestoßen, die nur deshalb überlebt hatte, weil die Zeit vergessen hatte, sie auszulöschen' (Hillaby 1964).

Tab. 3. Mikrobiologische Untersuchung von Lake Turkana-Wasser, Ugalli und getrocknetem Nilbarsch (1977)

	Lake-Turkana-Wasser Elmolo-Bucht	Nilbarsch (trocken)	Maismehl (Ugalli)
Kolonienzahl	1 ml Wasser: in Gelatine 20°C 205600 Gelatine verflüssigende Keime 4600 in Nähragar 20°C 292000	in gr Material: in Gelatine 20°C 4000 in Nähragar 20°C 56600	in gr Material: in Gelatine 20°C 12200 in Nähragar 20°C 17700
Coli-Titer	in 1,0 – 0,1 ml Wasser negativ	in 10,0 – 0,1 ml Suspension negativ	in 10,0 – 0,1 ml Suspension negativ
Coli-Zahl	in 1,0 und 0,5 ml Wasser negativ	in 1,0 und 0,5 ml Suspension negativ	in 1,0 und 0,5 ml Suspension negativ
Mikrobiologische Kultur (aerob und anaerob)	Clostridium bifermentans Sporenbildner der Subtilis-Mesentericus-Gruppe Achromobacter spec Bakterien der Pseudomonas-Gruppe Flavo-Bakterien anhaemolysierende Streptokokken Bakterien der Typhus-, Paratyphus-, Ruhr-, Enteritis-Gruppe wurden nicht nachgewiesen	Sporenbildner der Subtilis-Mesentericus-Gruppe Staphylococcus albus Aspergillus spec. vergrünende Streptokokken Entebacter cloacae Bakterien der Typhus-, Paratyphus-, Enteritis-Ruhr-Gruppe wurden nicht nachgewiesen	Clostridium ramosum (Sporenbildner der Subtilis-Mesentericus-Gruppe) Enterobacter agglomerans Penicillium spec. Bakterien der Typhus-; Paratyphus-, Enteritis-Ruhr-Gruppe wurden nicht nachgewiesen

Seit Graf Teleki die Elmolo 1888 entdeckte, hat die Zeit ihr Leben kaum verändert.

Die Elmolo-Siedlung auf der Insel Koran (Abb. 15, 16, 17, 18) macht einen unordentlichen und unsauberen Eindruck. Die wenigen Kuppelhütten stehen wahllos verstreut auf dem kahlen, steinigen Lavaboden, mit Laichkraut oder Palmenwedeln bedeckt und vom Winde zerfetzt. Zwischen den Hütten liegen Abfälle jeglicher Art: Reste von Netzen, Tonscherben, Fischgeräten, alte Blechbüchsen und sonstiges Inventar. Nördlich des Dorfes sammelt sich der Unrat zu einer Art Abfallhaufen mit Speiseresten, Fischkadavern, Krokodilschädeln, Schildkrötenpanzern und vielem anderen mehr.

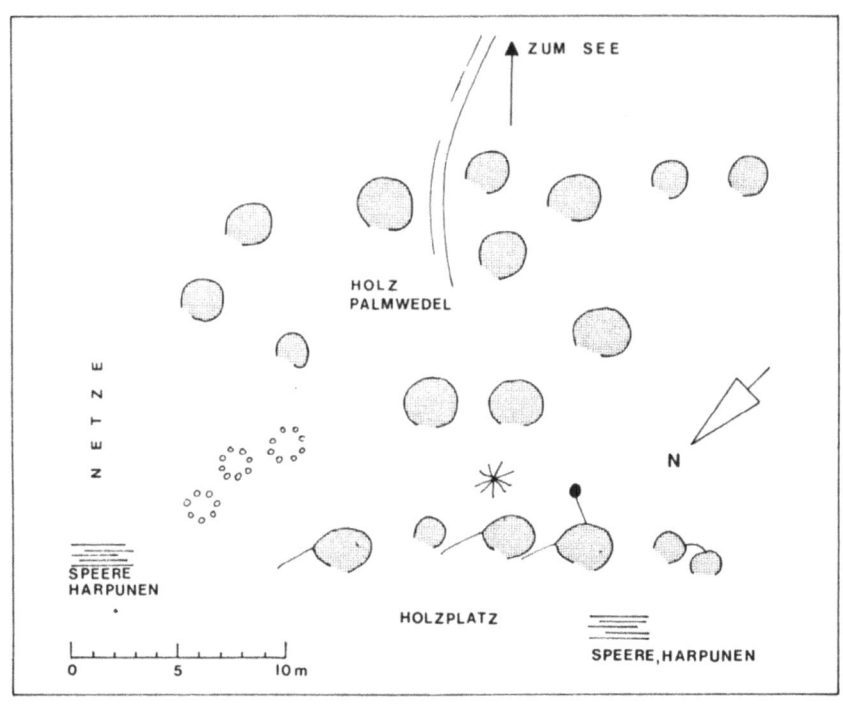

ZUM SEE

HOLZ
PALMWEDEL

N E T Z E

N

SPEERE
HARPUNEN

HOLZPLATZ

SPEERE, HARPUNEN

0   5   10 m

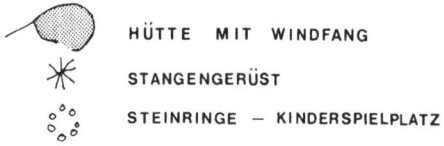

HÜTTE MIT WINDFANG

STANGENGERÜST

STEINRINGE — KINDERSPIELPLATZ

Abb. 15. Lageplan der Elmolo-Siedlung auf der Insel Koran.

Zwischen all dem spielt sich der Alltag der Elmolo ab. Die Männer sind mit dem Anfertigen und Ausbessern ihrer Jagdgeräte beschäftigt. Die Frauen (Abb. 19) hocken in Gruppen zusammen, fertigen Schmuck an, unterhalten sich und stillen ihre Kinder. Einzelne nehmen Fische aus (Abb. 20), machen Feuer zum Kochen und setzen die schmutzigen, verrosteten Blechbüchsen und Tontöpfe auf. Rauch erfüllt die kleinen Hütten, und der Fischgeruch steigert sich in der Mittagshitze ins Unerträgliche. Die Jüngeren gehen zum Trinken ans Ufer (Abb. 14) und benutzen gleichzeitig die Gelegenheit, ihre Fäkalien dort ins Wasser abzugeben, wo auch die Frauen ihr Kochwasser entnehmen. Die wenigen alten Leute hocken, schmutzig und zerñumpt, zusammengekauert in ihren kleinen Hütten, blind und unfähig sich zu bewegen (Abb. 9).

Die hygienischen Verhältnisse lassen sehr zu wünschen übrig. Daher ist es auch nicht überraschend, daß die mikrobiologischen Untersuchungen des Seewassers und des getrockneten Nilbarschfleisches sowie des Ugalli einen hohen Verunreinigungsgrad zeigen (Tab. 3).

Abb. 16. Die 2 Jahre alte Elmolo-Siedlung auf Koran besteht aus insgesamt 18 Hütten, deren 1 m hohe Eingänge einheitlich nach NW (windabgewandte Seite) gerichtet sind. Die Hütten, die je nach Größe von 1-7 Personen bewohnt werden, sind ca. 1,50 m hoch und zum Teil mit einem Windfang versehen.

Trotz der starken Verschmutzung war die Coliuntersuchung jedoch erstaunlicherweise negativ. Bakterien der Thyphus-, Parathyphus- und Ruhr-Enteritis-Gruppe konnten ebenfalls nicht nachgewiesen werden. Bei diesen Gruppen handelt es sich allerdings um nicht sehr resistente Bakterien, die in einer wässrigen Lösung durch Konkurrenzverhalten von anderen Bakterien sehr schnell aufgezehrt und vernichtet werden. Aufgrund der übrigen Ergebnisse können wir daher mit hoher Wahrscheinlichkeit davon ausgehen, daß diese pathogenen Bakterien ebenfalls vorkommen.

Die festgestellten Bakterien sind größtenteils nur fakultativ pathogen. Bei gutem Ernährungszustand der Bevölkerung richten sie keinen Schaden an. Unter besonderen Lebensbedingungen sowie bei Wunden, bei Augen- und Ohrenverletzungen können sie jedoch sehr schädlich wirken. Dies trifft besonders für 'Clostridium ramosum' zu, das bei der Untersuchung des Ugalli festgestellt wurde. Wird das Maismehl mit Seewasser angerührt und bleibt dann unter dem vorherrschenden trockenheißen Klima nur kurze Zeit stehen, so kommt es zu einer raschen und massiven Kolonienbildung von Clostridium ramosum. Der Genuß dieses infizierten Maismehlbreis führt dann nach kurzer Zeit zu Brechdurchfallerscheinungen. So kann das Ugalli, obwohl sehr nährstoffreich, zusammen mit Seewasser eine schädliche Wirkung haben.

Zur weiteren Ursachenanalyse der Knochendeformierungen wurde eine halb-quantitative Bestimmung von Calciumionen im Urin durchgeführt. Dabei ergab sich bei 10 untersuchten erwachsenen Elmolo eine Calciumausscheidung von 0 bis 25 mg/1. Dagegen wurden bei unseren Expeditionsteilnehmern Ausscheidungen zwischen 75 bis 100 mg/1 festgestellt (Tab. 4).

Der Verdacht einer Lueserkrankung, die mit ähnlichen Knochenveränderungen einhergehen kann wie sie bei den Elmolo auftreten, wurde eindeutig widerlegt. Bei

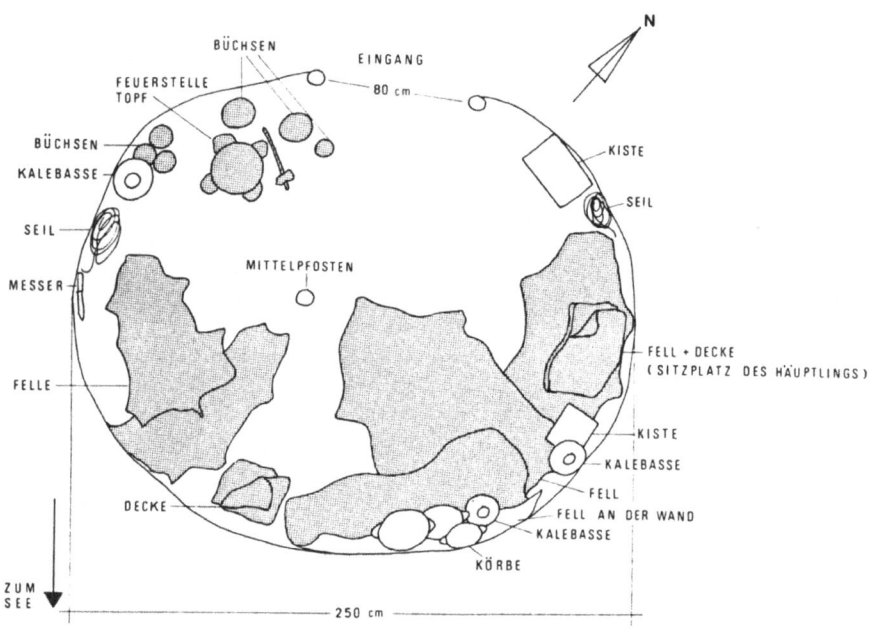

Abb. 17. Hüttengrundriß.

Tab. 4. Urin-, Lues- und Tuberkulosetest an verschiedenen Elmolo-Kollektiven (1977)

Untersuchung	Anzahl	Befund	%
Ca – Ausscheidung (Urin)	10	0 – 25 mg/l	100
Luestest	10	–	100
Tinetest Insel (Tbc)	20	+ –	10 90
Tinetest-Uferdorf (Tbc)	20	+ –	70 30

Abb. 18. Ein alltägliches Stilleben aus Tradition und Fortschritt: Schildkrötenpanzer und Tontopf neben Blechbüchse und Plastikflasche. Mangelnde Hygiene und Unordnung sind ein Charakteristikum der Elmolo-Siedlung.

der Untersuchung von 10 erwachsenen Elmolo mit einem serologischen Schnell-test konnten nur negative Werte gefunden werden (Tab. 4).

Ein auf der Insel durchgeführter Tine-Test auf Tuberkulose an 20 Elmolo ver-schiedenen Alters und Geschlechts zeigte bei 90% negative, bei 10% positive Wer-te. Dies besagt, daß 90% der Elmolo mit der Tuberkulose überhaupt noch nie in Berührung gekommen sind. Derselbe Test im Uferdorf Anderi durchgeführt, ergab jedoch bei 20 Elmolo verschiedenen Alters und Geschlechts 70% positive und 30% negative Werte. Durch die stärkere Kommunikation im Uferdorf ist ein wesentlich größerer Teil der Bevölkerung bereits mit der Tuberkulose konfrontiert worden und hat dadurch natürliche Abwehrkräfte entwickelt (Tab. 4).

Neben den Nahrungs- und Ernährungsgewohnheiten sowie den hygienischen Verhältnissen spielen für die Beinverbiegungen der Elmolo möglicherweise auch ihre Sitzgewohnheiten eine entscheidende Rolle. Sie nehmen stundenlang eine Hockstellung ein, wenn sie sich im Schattenbereich ihrer niedrigen Hütten aufhal-

Abb. 19. Diese Frau trägt noch die Selah, die traditionelle Kleidung der Elmolo-Frauen aus Palmfasern. Den Perlenkettenschmuck haben die Elmolo von den Samburu übernommen. Narbentätowierung ist sowohl bei Männern als auch bei Frauen üblich.

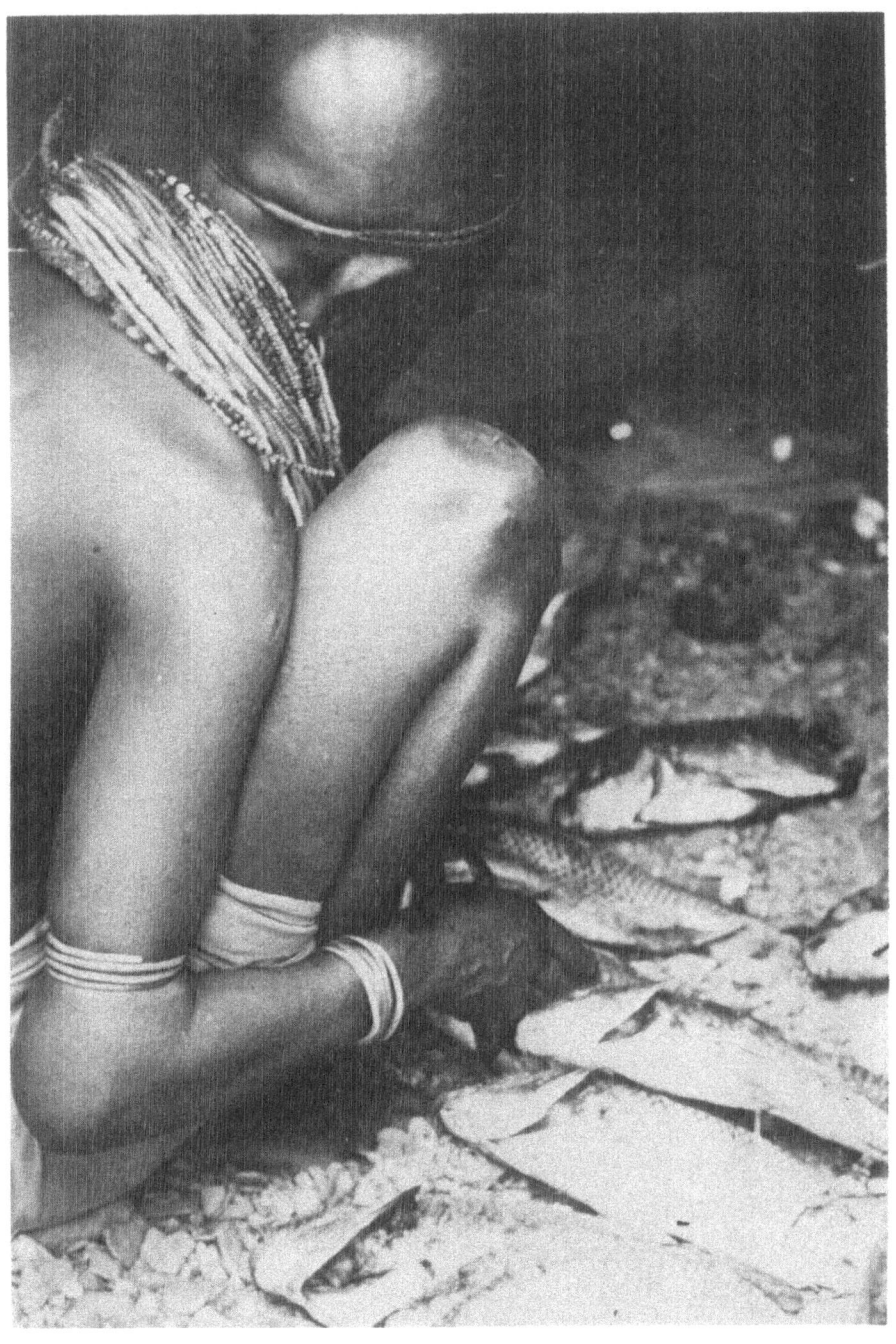

Abb. 20. Die Elmolo-Frauen verrichten fast alle 'Hausarbeiten' in Hockstellung, wodurch Knochendeformierungen hervorgerufen werden könnten.

Abb. 21. Das bei den Elmolo übliche Tragen von Kindern an der Körperseite beeinträchtigt die Körperhaltung und könnte Deformierungen am Skelettapparat verursachen.

ten (Abb. 20). Bei dieser Sitzart übt das Körpergewicht auf die Schienbeine einen gleichzeitigen Druck nach vorn und außen aus, wodurch die Ausprägung von O-Beinen begünstigt wird.

Verformungen der Wirbelsäule, die bei Frauen häufiger auftreten als bei Männern, lassen sich zumindest teilweise durch die stärkere statische Belastung beim Tragen ihrer Kinder an der Körperseite erklären (Abb. 21). Außerdem wird ihr Stoffwechsel während der Schwangerschaft und Stillzeit ungleich mehr beansprucht.

Darüber hinaus ist nicht auszuschließen, daß ein bereits vorliegender genetischer Faktor für die Knochenverformungen mitverantwortlich ist. In einer derartig kleinen Gemeinschaft ist Inzucht trotz postulierter Sippenexogamie nie ganz auszuschließen, was die Ausprägung bestimmter negativer Merkmale begünstigen kann.

Die Elmolo selbst bringen ihre Knochenerkrankungen in Zusammenhang mit dem Wasser, jedoch ohne weitere Kenntnis der Ursachen. Die Behandlung erfolgt mittels einer Wärmetherapie. Sie legen auf die schmerzenden Gelenke ein Stück Ziegenfell, in das vorher heiße Asche eingepackt wird, und pressen es so lange auf die Stellen, bis diese über und über mit Brandwunden bedeckt sind.

Aufgrund der bisherigen Untersuchungsergebnisse ist wahrscheinlich, daß es sich bei den Knochenveränderungen der Elmolo um eine Rachitis bzw. Spätrachitis handelt, die aus einer andauernden qualitativ mangelhaften Ernährung resultiert. Die auslösenden Ursachen sind eine zu geringe Zufuhr an Vitamin D, an Calcium sowie an Spurenelementen.

Vitamin D als Provitamin wird dem Körper hauptsächlich mit Milch und Milchprodukten zugeführt. Diese Nahrungsmittel fehlen den Elmolo gänzlich. Die durch ultraviolette Strahlung auf die Haut gebildete Menge an Vitamin D ist nur unzureichend (Wolf 1974, Kenntner/Pfannkuch 1976).

Knochenaufbau und Knochenumformung sind komplexe Stoffwechselvorgänge. Sie sind abhängig von einer ausreichenden Menge und Resorption von Calcium im Darm unter Mitwirkung von Vitamin D und dem Schilddrüsenhormon Calcitonin. Selbst wenn ausreichend Calcium mit der Nahrung aufgenommen wird, kann die Resorption aus dem Darm und der Einbau in die Knochensubstanz erst stattfinden, wenn Vitamin D als Vermittler für die beiden Stoffwechselvorgänge vorhanden ist. Fehlt dieses Vitamin, kann das Calcium in der Knochenmatrix nicht abgelagert werden, die Folgeerscheinung ist Rachitis. Im Blut des Menschen besteht ein konstanter Serum Calciumspiegel von 9,0 bis 11,0 mg %. Dieser Wert wird ständig durch das Hormon der Nebenschilddrüse (Parathormon) geregelt. Sinkt der Calciumspiegel unter 9 mg%, weil die Nahrung zu wenig Calcium enthält, der Vitamin D-Gehalt unzureichend ist und doch möglicherweise vorhandene Nieren- und Darmerkrankungen zuviel Calcium ausgeschieden wird so mobilisiert das Parathormon unverzüglich Calcium aus den Knochen, um den Serum-Calciumspiegel wieder zu normalisieren. Dies führt zwangsläufig zu einer Entkalkung des Knochens bzw. zu Spätrachitis (Rapoport 1973, Wolf 1974, Kenntner/Pfannkuch 1976). Ein derartiges Wirkungsgefüge kann aufgrund, unserer Untersuchungen für die Elmolo angenommen werden.

Tab. 5. Chemische Analyse von Ugalli im Vergleich zu europäischem Maismehl 1977

Substanz	Ugalli	Maismehl
Wasser	10,3%	12,0%
Fett	4,2%	2,8%
Mineralstoffe	1,1%	1,2%
Protein (N x 6,25)	8,4%	9,0%
Phosphor (P_2O_5)	0,5%	1,2%
Calcium (Ca)	0,02%	0,02%
Eisen (Fe)	0,002%	0,002%

Daraus ergeben sich folgende Möglichkeiten, den physischen Zustand der Elmolo zu verbessern: 1. Aufstockung der einseitigen Fischernährung, 2. Verbesserung der Trinkwasserqualität, 3. Änderung bestimmter Lebensgewohnheiten (Hygiene, Sitzgewohnheiten).

Außerdem kann durch die Verwendung von Ugalli die einseitige Fischernährung wesentlich bereichert werden. Chemische Analysen dieses Maismehls ergaben einen ausreichenden Gehalt an Nährstoffen (Tab. 5). Es ist jedoch darauf zu achten, daß Ugalli nicht mit Seewasser in Verbindung kommt. Qualitativ bessere Trinkwasserquellen müßten daher ausfindig gemacht werden.

Eine Verbesserung der Trinkwasserqualität könnte durch Erbohren von Grundwasser oder durch chemische Aufbereitung des entnommenen Seewassers erreicht werden. Bei der Seewasserentnahme wird jedoch darauf zu achten sein, daß die Elmolo nicht an derselben Stelle ihr Trinkwasser entnehmen, an der sie auch ihre Fäkalien in den See abgeben.

Aus humanitären Gründen ist eine Unterstützung der Elmolo auf schnellstem Wege zu fordern.

Für die Wissenschaft sind die Elmolo ein einzigartiger Fall zum Studium von Mensch-Umwelt-Beziehungen, über die wir bisher noch nicht allzu viel wissen. Weitere Untersuchungen sollten daher sobald als möglich erfolgen, da es um die Zukunft der Bevölkerung schlecht bestellt ist. Der 'kleinste Stamm Afrikas' wird in absehbarer Zeit verschwunden sein, denn es existieren heute nur noch etwa 230 Elmolo, davon 64 auf Koran.[1]

---

[1] In Anderi wurden 142 Elmolo gezählt. Berücksichtigt man, daß einige Personen zum Zeitpunkt der Erhebung abwesend waren bzw. in anderen Stammesverbänden leben (Mischehen), so dürfte die Zahl von 230 in etwa den Tatsachen entsprechen.

Problematisch sind die Altersangaben. Erwachsene wußten in den meisten Fällen ihr Alter gar nicht oder nur sehr ungenau anzugeben, sodaß Schätzungen und Korrekturen vorgenommen werden mußten.

Die Alterspyramide (Abb. 22) zeigt die für ein Entwicklungsland typische Pyramidenform, deren Basis jedoch nicht allzu breit ist, d.h. die Kinderzahl ist nicht sehr hoch. Die durchschnittliche Familiengröße beträgt 3,46 Personen; nur wenige Familien bestehen aus 5 und mehr Mitgliedern (Abb. 23). Dies besagt, daß rein statistisch der derzeitige Bevölkerungsstand nicht erhalten bleibt. Es wird in

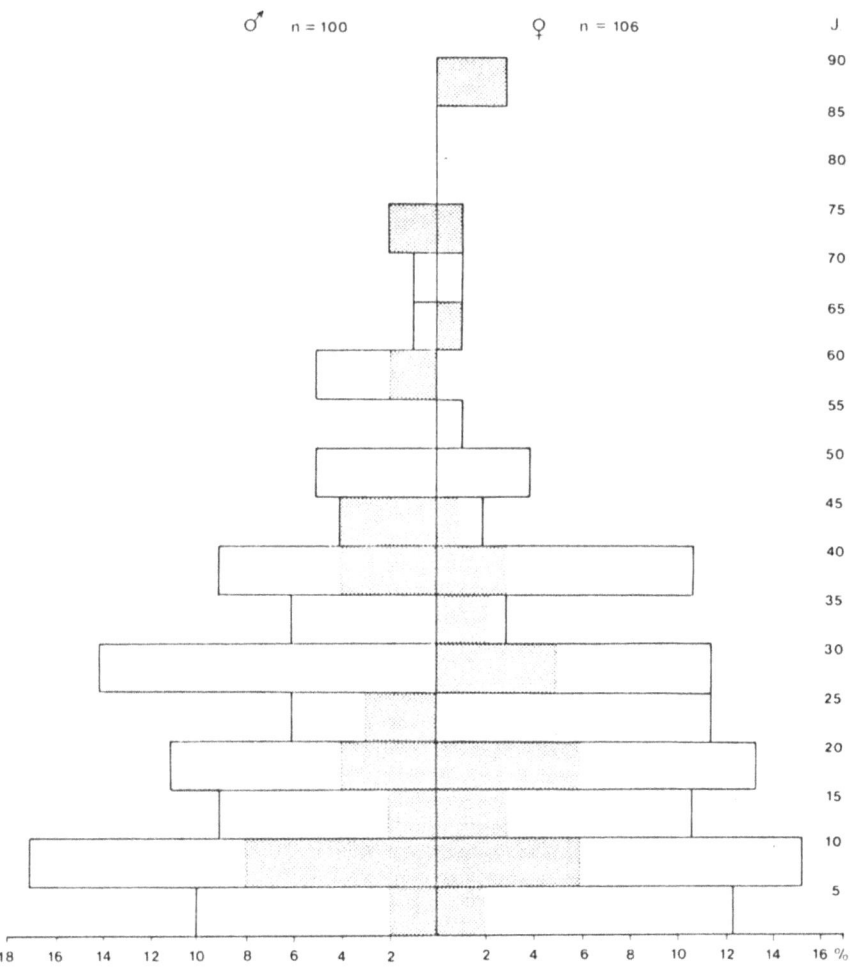

BEVÖLKERUNGSPYRAMIDE DER ELMOLO – GESAMTPOPULATION

DAVON : = INSELPOPULATION

Abb. 22. Alterspyramide.

189

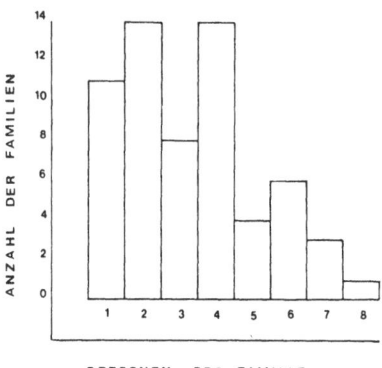

FAMILIENSTRUKTUR DER ELMOLO

GESAMTZAHL : 59 FAMILIEN
DURCHSCHNITT : 3,46 PERSONEN PRO FAMILIE

ANZAHL DER FAMILIEN

PERSONEN PRO FAMILIE

Abb. 23. Familienstruktur.

naher Zukunft eine Abnahme der Bevölkerung zu erwarten sein, vor allem auch wohl deshalb, weil der Absorbierungsprozeß fortschreitet. Junge Elmolo-Frauen werden zunehmend in andere Stammesverbände integriert, den Elmolo-Männern gehen dadurch mögliche Ehepartner verloren.

Ein besonders hoffnungsvolles Bild zeigt die Alterspyramide der 64 Elmolo von Koran (Abb. 22): Total unterrepräsentiert sind Männer im Alter zwischen 25 und 40 Jahren sowie Frauen im gebärfähigen Alter. Dementsprechend gering ist die Zahl der Kleinkinder und Babies. Nur ein junges Ehepaar konnte festgestellt werden und kein Fall von Schwangerschaft. Der Altersunterschied zwischen den Ehepartnern ist zum Teil groß: er beträgt 10 Jahre und mehr. Bei beiden Geschlechtern fehlen die höheren Jahrgänge, was auf eine frühe Sterblichkeit und damit niedrige mittlere Lebenserwartung schließen läßt. Die wenigen ganz alten Leute machten alle einen sehr armseligen Eindruck.

Das Fehlen des aktivsten Bevölkerungsteiles weist Koran als Auswanderungsgebiet aus. Die meisten jungen Leute sind, wie in der Alterspyramide für die Gesamtpopulation zum Ausdruck kommt, ins Uferdorf Anderi gezogen; die übrigen werden folgen oder als überalterte Bevölkerung in absehbarer Zeit aussterben.

In wenigen Jahren werden die Elmolo-Inseln verlassen sein. Zwar wird die Population als solche noch eine Zeitlang weiterbestehen, aber nicht mehr in ihrer ursprünglichen Natur- und Kulturumwelt und auch nur noch solange der Vermischungsprozeß nicht alle Elmolo erfaßt hat.

Zusammenfassung

Die Elmolo-Population am Lake Turkana befindet sich in einem außerordentlich schlechten physischen Zustand. Deformationen am Skelettapparat kommen in auffälliger Häufigkeit vor. Neben populationsspezifischen Verhaltensweisen sind vor allem die ökologischen Bedingungen ihres Lebensraumes, die Ernährungsverhältnisse und Hygiene dafür verantwortlich zu machen.

Die primäre Ursache ist wohl in der einseitigen und defizitären Nahrung der Elmolo zu suchen. Fisch- und Wasseranalysen haben ergeben, daß in der Nahrung der Elmolo eine ungünstige Mineralstoffkonstellation vorherrscht, die vor allem in einem Ca-Mangel zum Ausdruck kommt. Demgegenüber sind hohe Konzentrationen an Fluor und Ammoniak feststellbar. Der Verdacht auf Fluorosis kann jedoch nicht bestätigt werden. Es ist vielmehr wahrscheinlich, daß es sich bei den Knochendeformierungen um eine durch Mineralstoffmangel bedingte Rachitis oder Spätrachitis handelt. Neben dem Mineralstoffmangel sind die schlechten hygienischen Verhältnisse sowie bestimmte Lebensgewohnheiten (stundenlange Hockstellung, Tragen von Lasten an der Körperseite) als möglicherweise mitverantwortliche Faktoren anzusprechen.

Aufgrund der geringen Individuenzahl und der zunehmenden Abwanderung nach Loiyangalani wird in absehbarer Zeit eine anthropologisch und medizinisch hochinteressante Population und eine einzigartige noch weitgehend unberührte und kaum bekannte Kultur verschwunden sein. Sowohl aus humanitären als auch aus wissenschaftlichen Gründen ergibt sich daher die dringende Notwendigkeit, Möglichkeiten zu erarbeiten, um die kleine Population zu erhalten.

Literatur

Adamson, G. 1969. Safari meines Lebens, Hamburg.

Adamson, J. 1960. Born free, London, Glasgow.

Adamson, J. 1967. The Peoples of Kenya, London.

Chlud, K. 1973. Rechtliches zur Dosierung von Fluorpräparaten, Wochz. prakt. Med. H. 43/73 Jahrgang 23.

Dyson, W.S. & Fuchs, V.E. 1937. The Elmolo, *Journ. Roy. Anthr. Inst.* 67, S. 327-338.

Franke, J. 1973. Knochenfluorose, Wochz. für prakt. Med., H. 43/73, Jahrgang 23.

Fuchs, V.E. 1935. The Lake Rudolf Rift Valley Expedition, 1934, *Geogr. Journ.* 86: 114-142.

Graham, A. & Beard, P. 1973. Eyelids of Morning. The Mingled Destinies of Crocodiles and Men. New York.

Henderson, J. 1958. Family Portrait of a complete African Tribe, Sunday Times, 19. Okt. 1958, S. 7.

Hillaby, J. 1964. Die Reise zum Jadesee. Berlin, Frankfurt.

Höhnel, L.R. v. 1892. Zum Rudolph-See und Stephanie-See. Die Forschungsreise des Grafen Samuel Teleki in Ost-Äquatorial-Afrika. 1887-1888, Wien.

Harvey, G. 1966. The People of Lake-Rudolf. *Africana* 2, No. 8, S. 27-29.

Hory, A.C. 1911. Lake Rudolph. *Journ. East Afr. Nat. Hist. Soc.* 12, S. 47-52.

Kenntner, G. & Pfannkuch, K.H. 1976. Elmolos. Ein Volksstamm stirbt. *Bild der Wissenschaft* 12, s. 42-50.

Lloyd/Jones, W. 1925. Hàvash. Frontier Adventures in Kenya, Bristol.

Matthiessen, P. & Porter, E. 1973. Der Baum der Schöpfung. Wien, Zürich, München.

McDougall, J. 1974. Another Twilight at Lake Rudolf. Of a People and their Life Style. *Africana* 5, No. 7, S. 21-24.

Nestmann, L. 1974. Human Development in its Relations to ecological Conditions. *Geoforum* 18, S. 7-18.

Rapoport, S.M. 1973. Med. Biochemie, 5. Aufl.

Robbins, L.H. 1974. The Lothagam Site. Michigan State University, Anthr. Ser. 1, No. 2.

Spencer, P. 1973. Nomads in Alliance. London.

Schwidetzki, I. 1962. Die neue Rassenkunde. Stuttgart.

Srikanta, S.G. 1965. Metabolic Studies in Skeletal-Fluorisis Clin. Sci. 28, 477.

Vagnby, B. & Jacobs, A.H. 1974. Kenya traditional housing of the Elmolo. Ekisties 38, No. 227, S. 240-243.

Willock, C. 1974. Das Afrikanische Riftvalley. Amsterdam.

Wolf, H. 1974. Rachitisprophylaxe. Hundt-Verlag, Hattingen.

Worthington, E.B. 1932. The Lakes of Kenya and Uganda, investigated by the Cambridge Expedition 1930-31, *Geogr. Journ.* 79, S. 275-297.

Abbildungsnachweis:

Abb. 1, 2, 3, 4, 5, 9, 10, 11, 13, 14, 15, 17, 20, (22), (23)	H. Ohliger
Abb. 7, 12, 16, 19, (22), (23)	E. Ludwig
Abb. 6, 8, 18, 21, (22), (23)	G. Kenntner

Address of the Author's:
Prof. Dr. G. Kenntner und
E. Ludwig, Geographisches Institut der Universität des Saarlandes
66 Saarbrücken, FRG

OPERATIONAL REMOTE SENSING SYSTEMS FOR REGIONAL PLANNING AND ENVIRONMENTAL MONITORING

S. Schneider

The growing experience gained from a variety of civil applications of remote sensing techniques during the last decade have demonstrated a more efficient systematic and cheaper approach to environmental data collection and field survey planning. There has been opened a broader view of the interrelationships between various natural phenomena and human activities. But the fuller utilization of remote sensing technology by many public services and managements, may be expected to bring even more important achievements during the next decades through the qualitative and quantitative improvement of the remotely sensed information and the increased effectiveness of its procurement. This can be achieved through the agency of quick, large to medium scale, multi-sensor observation of special features within small areas by aircraft, or by less detailed, small to global scale but repetitive and synoptic surveying by satellites.

As experience has shown, remote sensing provides the data for area-information in many fields of practical ecological application for regional planning and environmental monitoring, such as for example:
– Land use (inventory and planning), land-use changes, land-use conflicts, control of planning achievement,
– agricultural crop identification,
– plant diseases (crop and forest protection),
– soil erosion,
– Disaster assessment and prediction (volcanoes, earthquakes, landslides, inundations),
– Global weather mapping, regional climatic conditions (temperature),
– Surveying of amphibian zones,
– Water pollution monitoring,
– Air pollution monitoring,
  'Remote sensing systems have to offer various levels of the sensing capabilities, various degrees of reliability, various stages of information processing, dependent on the areas of application and on the degree of development from experimental to operational utilization.' (Schanda, 1976). The first example should be represented by land-use inventory. It is widely accepted, in terms of cost, time and accuracy, that the most favorable method of land-use mapping is by means of remote sensing.

In Central and western Europe the conventional official land-use data are limited to certain traditional land-use classes reflecting the cultural landscape of several decades ago. Consequently they are based on expanding units. In other words, the territories of our communities as the smallest statistical reference units are increasing in size through administrative centralization.

In Germany there existed in 1961 a total of 24 503 communities; in 1975 the total number of communities was only 10 914 without any change of the total spatial dimension of the state area. The official statistical information based on communities is therefore no longer so detailed as before.

Remote sensing data are complementary to those statistical data which present detailed information such as the areal use of types of buildings, the number of floors and the floor space.

Moreover remote sensing data expand the available information to cover ecological, socio-economic and environmental problems such as:
the monitoring of air and water-pollution, of the different sorts of fallow-land, of the open spaces in the cities or of the progress in reclaiming land in opencast-mining districts. Our experience proves that the task of land-use survey by photo-interpretation techniques could be solved with an interpretation accuracy about 90 to 95 % (provided: the right scale).

The systematical development of airphoto interpretation in combination with computerized data handling has led to the use of aerial photography being favoured for land-use analysis as a first step to a regional information system.

In relation to the conventional cadastral classification actual remote sensing data have presented more exact results of areal identification and dimension of up to 25-30 %.

Moreover, the land use data of our test field, the Rhine-Neckar region, have been digitized in a very short time and recorded in a regional handbook of land use statistics.

For the first time we were able to present a complete digitized information on the land use structure of the non built-up areas of the Rhine-Neckar region in the scale 1:50 000 resp. 1:25 000.

Because of the world-wide coverage by Landsat imagery, several countries have used this multispectral data material for an inventory of their territory and monitoring land use changes. The problems of rectification and mounting have been solved.

The Federal Department for National Planning has initiated a research project to examine the feasibility of transposing the Landsat data into a thematic map of land-use for planning purpose in the scale of 1:200 000. (cf. fig. 1)

The single test areas in the Upper Rhine Valley are enlarged to the interpretation scale of 1:50 000; the selected features are generalized to the final rectified map in the scale of 1:200 000.

The land-use categories and features are classified by a digital image handling system on a screen*; although it may be difficult in some instances to separate

* Interactive Digital Image Processing System (*DIBIAS*) of the German Aerospace Research Center. (DFVLR); object classification: Institute for Planning Data (IfP)

194

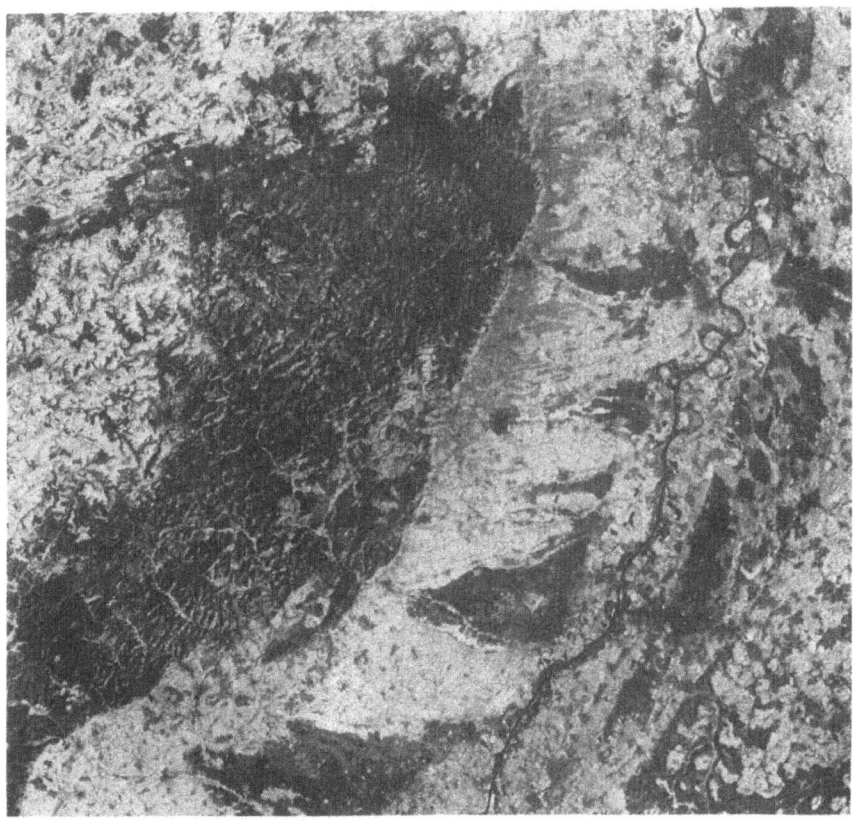

Fig. 1. Rectified sheet of the map CC 7110 (Mannheim), mounted of two LANDSAT-scenes of Aug. 9, 1975 and April 18, 1976.

real land use categories from signals without additional evidence, the results of the object-classification are promising. We hope this map will be the first sheet of a new thematic series of land use map.

Though Landsat 1 and 2-imagery cannot present thermal conditions (Landsat 3-march 78, has a thermal channel), it is possible to compare the distribution of haze and clouds over a large area, such as the Upper Rhine Valley by NOAA-imagery; cf. fig. 2, 3; it may be advisable to study multispectral satellite imagery. Over the Ruhr District a haze cover had been detected after filtering the four Landsat spectral images by a multispectral viewer.

The same problem, the influence of haze cover and clouds over large industrial plants along the adjacent residential areas leads back to the problems of city climate**. The temperatures of the open areas, their sequence in the course of a

** SVR (Siedlungsverband Ruhrkohlenbezirk) Essen, 1977 Geograph. Inst. I, Univ. Freiburg

195

Fig. 2. LANDSAT-1 imagery of the Upper Rhine Valley of Sept. 21, 1972. NASA-ERTS-E 1060-09552-6.

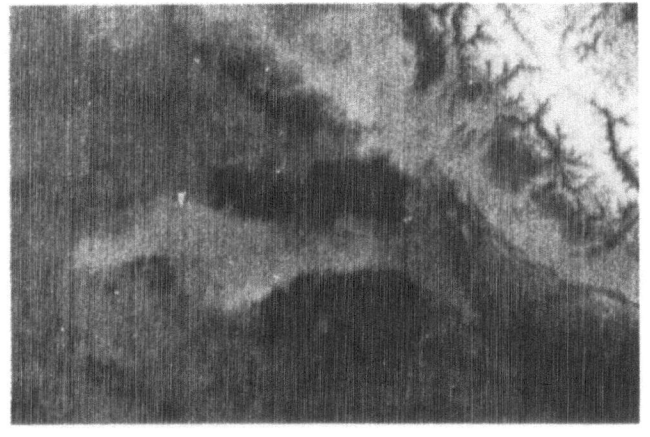

Fig. 3. JR. Imagery of the NOAA-Weather Satellite during an inversion on Febr. 7 (a), 9 (b) and 10 (c) of the Upper Rhine Valley. Photo Sternwarte Bochum, 1976.

197

night and their registration by a thermography offer the best means of attacking this problem.

Studies on the thermal conditions in the area of the Main Basin near Frankfurt and in the Rhine-Neckar Region near Mannheim led to the detection of fresh air corridors which are to be kept free of forests or housing.

In future it will no longer be possible changing the use of a larger area without monitoring the changes of the environmental conditions with the help of remote sensing technology at the same time.

Another problem in connection with the land-use inventory is the monitoring of land-use conflicts, such as the location of big chemical plants besides of recreation grounds or housing projects or between gravel-pits and water protection zones. This last problem has been studied in the Upper Rhine Valley where a large series of gravelpits follows the course of the river. Some of these gravelpits have been transformed into waste dumps.

Because of the delicate conditions for obtaining drinkingwater, the relation between the groundwater-level and the water level in the gravel pits has been studied last years using aerial photography and Landsat imagery. (cf. fig. 4)

The monitoring of recreational functions within the gravel-pit zone of the valley by color photography and by thermography has revealed the existence of potential land-use conflicts. It is just in the neighbourhood of urban agglomerations that the water in the gravel-pits and old river branches is the subject of considerable conflict.

The successful development of infrared technology — photography and radiometry — has led to its application in thermal studies of different planning projects. Nevertheless it must be admitted that some data obtained from a single sortie have been applied prematurely and too enthusiastically; and they have led to misinterpretation and wrong actions.

Two thermal infrared images taken within a period of several hours may yield completely inverse results. Therefore the success of infrared measuring and imaging depends to a high degree on the point in time and the meteorological conditions.

Several planning boards of the Ruhr-District, the Frankfurt-Area and the Rhine-Neckar-District have been encouraged to study this problem. The most interesting results were presented in the polycentric Ruhr-District with several large cities and between them buffer zones (green belts) which have so far been preserved from development (housebuilding). These buffer zones represent predominantly farm land with small woodland areas, public parks, cemeteries, allotments, playing and sports grounds in between. With the foundation of the Ruhr planning board in 1920, this board was made responsible inter alia for the configuration and limitation of green areas and zones, which should filter and renew the air of the neighbouring housing estates.

Even noise can be reduced by a corresponding configuration of the green areas. Because of its close neighbourhood to the residential areas, this regional green zone system is to a large extent suitable as a local recreation area. The thermo-

Fig. 4. Excavations and old river branches in the Rhine valley. Evaluation of a satellite picture (LANDSAT -1) of May, 12, 1933.

graphs made by line-scanner during day and night and different seasons present clearly the regional green zones in coloured equidensites. This gives an actual presentation of the differences in temperature around the areas covered with buildings of differing density as well as for the relatively sharp demarcation line around the built-up areas. The thermographs of the central Ruhr-District are an impressive proof of the fact that continuous open areas (green belts) have a cooling effect on cities. Generally speaking, the infrared themographies correspond with the measured city climate; the surrounding temperature also rises with an increasing density of housing.

The last example is taken from our research on monitoring and estimating water quality. We have models of an operational remote sensing system for different types of rivers in industrial districts. These data should help to optimize the location of power plants, industries, housing projects and recreation grounds. We have chosen the small Saar river, the Upper Rhine stream and the lower Elbe tidal estuary near Hamburg. During the flights over the rivers, we used a helicopter and a small one-engine aircraft. The helicopter equipped with an infrared radiometer for measuring the radiation-temperature above the water surface was flown in a longitudinal axis along the narrow Saar river and in 38 transverse axes over the Upper Rhine. Over small rivers a single central temperature curve may be sufficient for representing the temperature above the water surface. Over streams and estuaries it will be necessary to get the data of the water surface temperature in several curves, for instance in three curves along the centre of the river and along both sides. (cf. fig. 5) This situation is shown in the diagram of the Rhine near Mannheim-Ludwigshafen. Another form of presentation the temperature of the water surface would be the continuous print-out of scanned computerized data; in our sample one symbol represents 16 measured data in the Lower Elbe tidal river after the so-called Thermocomp-method. (cf. fig. 6) The radiation temperature-profiles across the river as well as the temperature imagery won by linescanning demonstrate that the distribution of temperature over the width of a big river may differ considerably.

The conventional methods of punctured measuring the water surface-temperature are not sufficient in comparison with the areal overview given by remote sensing methods. All essential discharges into the Saar river as well as into the Rine river have been detected. In both cases we had to combine the results of multispectral photography and of infrared scanning. In one case of a big cellulose plant the heavy polluted and heated discharge plume had been represented by both sensors; colour camera and infrared scanner.

The form and spread of the pollution plumes depend on the water quantity, the flow velocity and the temperature, so that the discharge plumes of the Saar and the Rhine have been usually quite different. (cf. fig. 7)

After flowing for a short time the plumes of the small Saar cover the whole river surface, the plumes of the Rhine are usually pressed along the banks as small and long bands extending in some cases more than 30 km in length. The mid-stream surface temperature of the Rhine is seldom influenced by the heated

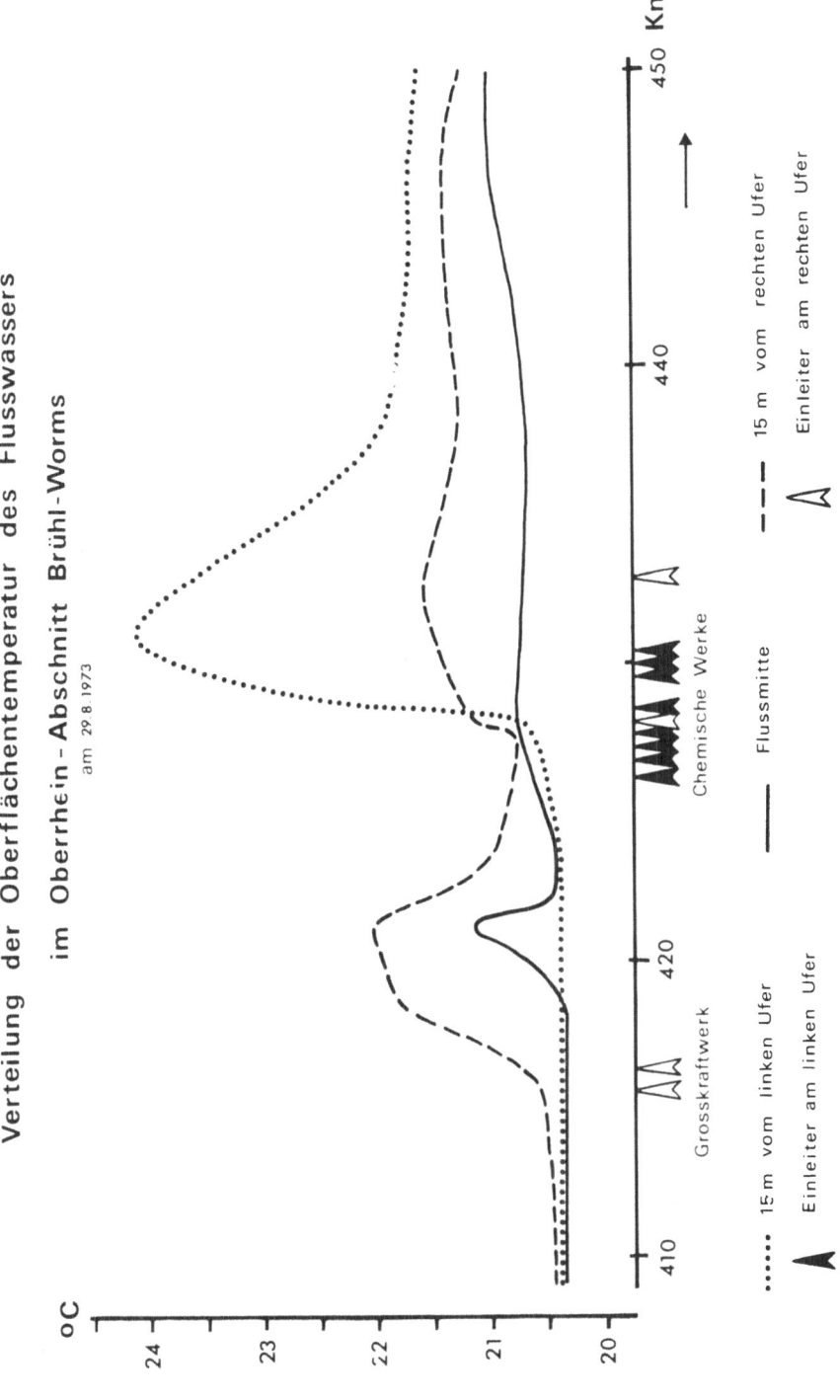

Fig. 5. Distribution of the surface water-temperature within the Upper-Rhine section Brühl-Worms near Mannheim - Ludwigshafen; ... 15 m from the left bank; — centre of the river; ... 15 m from the right bank; discharge of the left bank; discharge at the right bank.

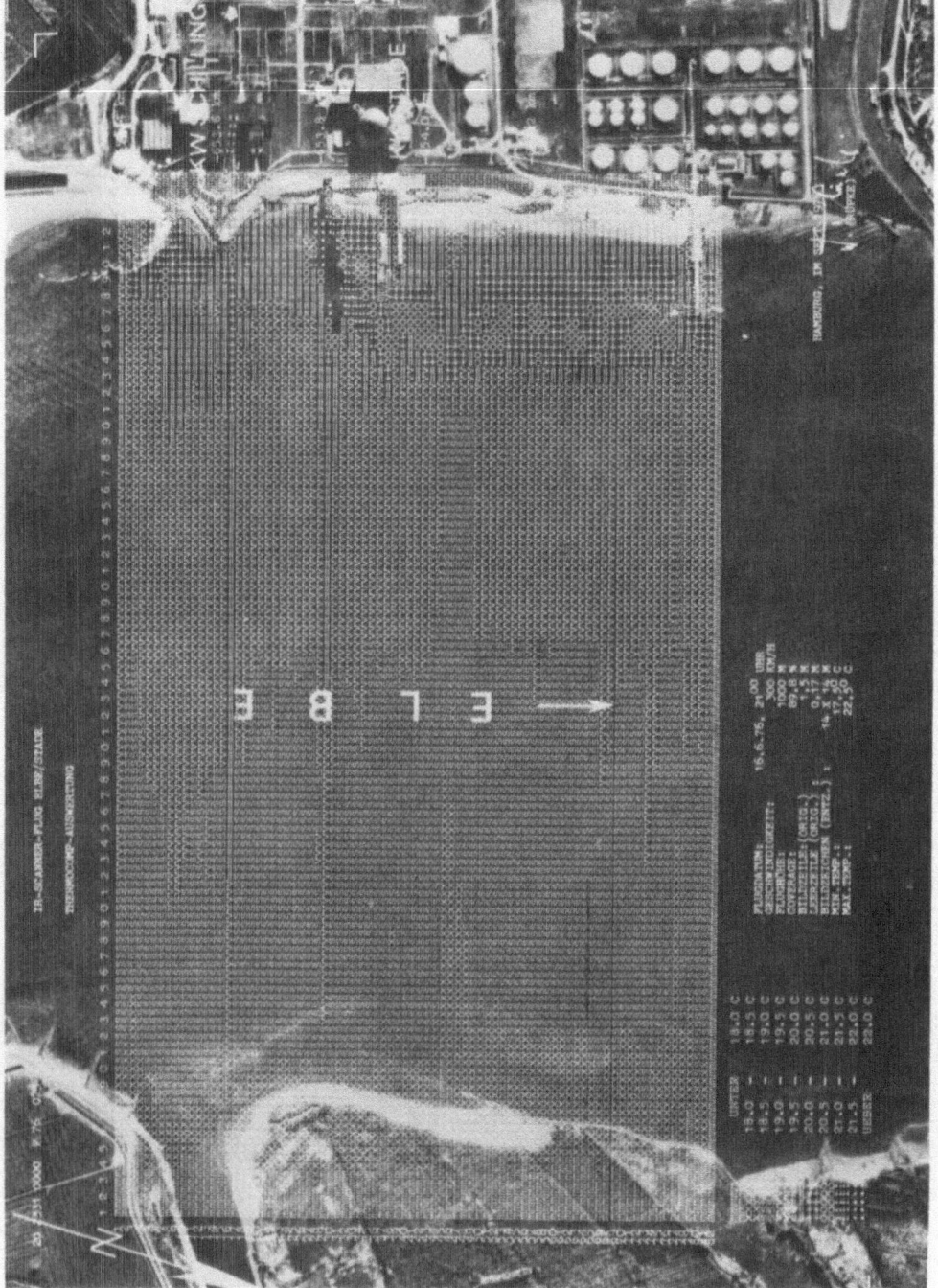

Fig. 6 Combination of aerial vertical photography and line scanner data over the tidal river of the lower Elbe near Stade. Heated discharge plumes of the conventional power plant Schilling and the atomic power plant Stade. The advantage of this picture is the digital representation

Fig. 7. Mouth of the Neckar into the Rhine river. Thermal (IR) imagery. The heated water of the Neckar covers in clouds the right bank of the Rhine. Two discharge plumes of the chemical works are running along the left bank of the Rhine. Photo: Hansa-Luftbild, Münster.

203

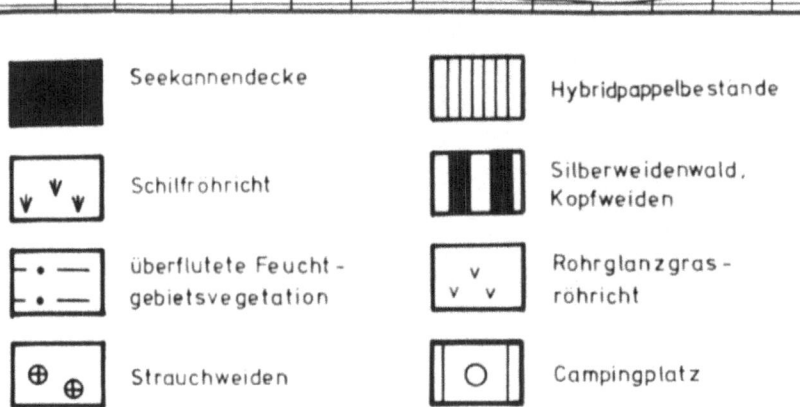

■	Seekannendecke	▥	Hybridpappelbestande
↓ ↓ ↓	Schilfrohricht	▮	Silberweidenwald, Kopfweiden
⋯	überflutete Feucht-gebietsvegetation	v v	Rohrglanzgras-rohricht
⊕ ⊕	Strauchweiden	○	Campingplatz

Fig. 8. Nature conservancy area 'Ludwigs insel'; different plant communities (G. Philippi).

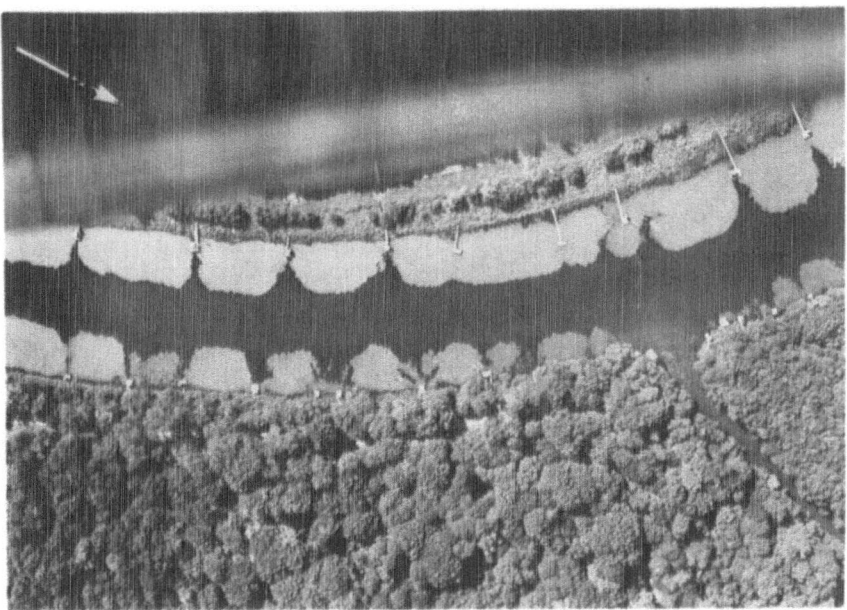

Fig. 9. Eutropnication on the banks of the Old Rhine near Wörth (Pfalz). Aquatic plants (Nuphar luteum and Trapa natans) are indicating the degree of trophication (III). Photo: G. Philippi, Karlsruhe, 1974; Freig.: Reg. Präs. Nordbaden, Nr 0/5486 v. 20.3.75.

discharge plumes near the banks, which means that the water surface temperature of big rivers in industrial and often densely populated regions is usually quite different in the zone close to the bank and in mid-stream.

An overheating of the river water would probably increase the oxygen demand of fish and other organism. An overheating of the Rhine would probably result in the danger that the biological balance in the water would be severely disturbed, causing a loss of the remaining selfpurifying power of the river. Ships would be troubled by the heavier development of mist. The recreational value of the river landscape would be lowered by increasing turbidity and visible algae development.

In extreme cases, mass formation of algae leads to intensive decomposition processes with decay and odour problems. Dangerous influences on the production, preparation and distribution of drinking water from surface water can be predicted, although not yet quantitatively determined.

The water quality of the small Saar river could be characterized by interpretation of the existence or non-existence of aquatic plants and plant-communities as indicators of pollution. In the case of the canalized waterway of the Upper Rhine the observation of aquatic plants must be confined to the zone of old branches and tributaries – 6-8 km of width – where the interpretation is based on certain plant communities as indicators of eutrophication. (cf. fig. 8)

It is precisely this wetland zone of old waters and tributaries, which is of great natural interest, which is now more endangered by human activities than the waterway itself. (cf. fig. 9) We identified more than 30 discharge points in the Upper Rhine Valley, where the sewage flew into the old waters and not immediately into the main stream. The land use conflicts arising in this zone include industrial location, location of power plants, refuse pits, pipe lines, gravel pits as against recreation and sporting grounds, forestry, agriculture and areas of nature conservation.

Because of the many possibilities of land-use conflicts in this wetland zone, a landscape inventory by airphoto interpretation has been carried out. First of all, infrared colour aerial photography, combined with ground checking, gives clear differentiations of plant societies.

Infrared represents smallest waters as well as waterlines, eutrophication near river banks, detergents, refuse pits, big industrial plants in spite of the mist and smoke in the valley.

As long as experiments and experiences of water turbidity and colour must be continued, we prefer the indirect way of water quality interpretation by studying the plant societies, espec. the hydroflora.

All data received by remote sensors and their digital picture processing such as the type of water, characteristics of flow, sort, source and colour of discharge plumes, distribution of surface temperature as well as the kind and degree of eutrophication make it evident that remote sensing methods can be successfully used for environmental monitoring and planning.

First results of remote sensing application are − the map on the water quality of the river Saar − maps on land-use conflicts in the valley-zone of the Upper Rhine and − temperature-maps of the tidal river Lower Elbe.

Using remote sensing methods, we have extreme valuable techniques for viewing the whole, regardless of how large it may be. If our objective is general information over very large areas, we have at our disposal satellite data. At the other extreme, we can observe phenomena just a few centimeters in size with very large scale aerial photography. It is of extreme importance that the right technique to the level of analysis and decision-making should be applied. To realize the full potential of remote sensing earth observation systems, these systems must move out of the experimental stage into an operational mode for ecological planning purposes on all scales.

Literature

Fezer, F. 1975. Infrarotbilder für die ökologische Standortbewertung und für den Städtebau. In: 'Erderkundung', Köln-Porz.

Gierloff-Emden, H.G. 1961. Luftbild und Küstengeographie am Beispiel der deutschen Nordseeküste. = Landeskundliche Luftbildauswertung, H. 4. Bad Godesberg.

Hassenpflug, W. & G. Richter 1972. Formen und Wirkungen der Bodenabspülung und -verwehung im Luftbild. = Landeskundliche Luftbildauswertung, H. lo. Bonn-Bad Godesberg.

Institut für Planungsdaten 1977. Realnutzungskartierung und Flächenbilanzierung im Bereich des Raumordnungsverbandes Rhein-Neckar. Offenbach (MS-Druck).

Kroesch, V. 1975. Wärmebelastung von Fließgewässern und ihre Erfassung durch Infrarot-Sensoren. In: 'Erderkundung', Köln-Porz.

Messerschmitt-Bölkow-Blohm 1973. Untersuchung des Nutzens der Fernerkundung der Erde mit den Mitteln der Weltraumtechnik.

Müller, Dietrich O. 1976. Verkehrserschließung und städtebauliche Entwicklung im Berliner Westen 1890-1975. 'Die Erde', 107 Jg. Heft 2-3.

Nithack, J. 1975. Landnutzungskartierung aus Landsat-1 und Skylab-Aufnahmen mit konventionellen Methoden. In: 'Erderkundung', Köln-Porz.

Regionale Planungsgemeinschaft Untermain 1970. Lufthygienisch meteorologische Modelluntersuchung in der Region Untermain. Arbeitsbericht 1970. Frankfurt a. Main.

Regionale Planungsgemeinschaft Untermain 1972. Lufthygienisch-meteorologische Modelluntersuchung in der Region Untermain. 3. Arbeitsbericht 1972 Infrarot-Thermographie. Frankfurt am Main.

Schanda, E. et el. 1976. Remote sensing for environmental studies, Berlin.

Schmidt-Kraepelin, E. & S. Schneider 1965. Luftbildinterpretation in der Agrarlandschaft und Beispiele ihrer Anwendung aus dem Lande Nordrhein-Westfalen. = Landeskundliche Luftbildauswertung, H. 7. Bad Godesberg.

Schneider, S. 1957. Braunkohlebergbau über Tage im Luftbild, dargestellt am Beispiel des Kölner Braunkohlereviers. = Landeskundliche Luftbildauswertung, H. 2. Remagen/Rh.

Schneider, S. et el. 1974. Gewässerüberwachung durch Fernerkundung. – Die mittlere Saar. Landeskundliche Luftbildauswertung, H. 12.

Schneider, S. et el. 1977. Gewässerüberwachung durch Fernerkundung – Der mittlere Oberrhein. = Landeskundliche Luftbildauswertung, H. 13.

Schneider, S. 1975. Remote Sensing and regional planning. In: Proceedings of the Seminar on regional planning cartography. Enschede, Niederlande.

Siedlungsverband Ruhrkohlenbezirk (SVR) 1972. Luftaufnahmen. Wärme-Infrarotmessungen als neue Informationsquelle für Planungszwecke. Essen.

Siedlungsverband Ruhrkohlenbezirk (SVR) 1975. Luftaufnahmen II. Auswertungen für Stadtplanung, Regionalplanung, Umweltschutz. Essen.

Troll, Carl 1966. Luftbildforschung und landeskundliche Forschung. = Erdkundliches Wissen, H. 12. Wiesbaden.

v. Hesler, A. 1975. Regionalplanung und Fernerkundung. In: 'Erderkundung', Köln-Porz.

Weischet, W. 1975. Stadtklimatologische Konsequenzen von Linescanner-Aufnahmen der Oberflächentemperaturen im Tagesgang (Beispiel Freiburg i. Br.). In: 'Erderkundung', Köln-Porz.

Adress of the Author:
Prof. Dr. Sigfrid Schneider
Kastanienweg 5
53 Bonn-Bad Godesberg, FRG

BASIC ECOLOGICAL CONCEPTS AND URBAN ECOLOGICAL SYSTEMS

Paul Müller

A representation of 'Basic Ecological Concepts' implies a thorough study of autecological, synecological and ecosystemic findings and discussions with research teams. Because of the limited time, however, I will confine myself to those whose paradigmatic consideration and methodological approach are of importance for the evaluation of urban systems. 'Urban Ecological Systems' would also deserve a broader treatment. Phylogenetically speaking, we have always faced the problem of bringing the environments which we have designed and shaped into harmony with our evolution-genetic inheritance and our ecological abilities. The institutions and environments which we have created are not the best ones imaginable, but only good enough to guarantee 'living together'. For this reason I want to include extensive ecological planning concepts only where a deliberately reductionistic standpoint has to be questioned. On the other hand, I will take pains to point out new research approaches as well as gaps in research up to the present, as far as through these a better ecological evaluation of areas and modification in their use seem possible. Principally, however, I do not want to evoke the impression that everything can be planned or should be planned.

1. Ecosystems, Burdening Capacity and Stability

A knowledge of the burdening capacity of ecosystems is a basic prerequisite for any kind of planning which takes the environment into account. But the 'burdening capacity' as well as the 'stability' of ecosystems can only be defined reasonably with regard to certain relationships (Fränzle 1978, Müller 1977).

Just like other systems, ecosystems can de differentiated from their surroundings in that their elements are more closely related to each other than to their environment (Abbott & Van Ness 1976). Therefore, three parameters are of fundamental importance for the description of ecosystemic structures: a. linkage type of the organisms (takes into consideration the kind of functional relationships; ecological niche) b. density of connections (as a measure for the webbing, takes into account the number of in- and outputs of the system's elements; food-web) c. type of webbing (describes the ratio between the in- and outputs of an element, of a system and those outside the system).

These three parameters demonstrate at the same time that as an example bioindication using wild species is possible in three fields (Müller 1976, 1978): a. chorological indication via surveys of occurrence, b. population-ecological indication by means of studying population fluctuations, c. biochemical-physiological indication through observation of the organisms' reaction, e.g. in relation to their intake of pollutants (accumulation indicators) and position within the food-web.

Since the elements of ecosystems are organisms with a history of development in some cases thousands of years old, it makes sense to speak of the genetic and ecological structure of an ecosystem. In hierachical order according to their adaptation to exogenous factors, ecosystems are open systems, because they transfer matter as well as energy to their surroundings.

As with thermodynamics, the conceptual state of balance plays an important role with ecosystems. Ecosystems are characterized by internal and external states of balance. The stability of the relations between elements determines the elasticity of the system's structure (Margalef 1975). The range of stability of an ecosystem therefore can be described in terms of the number of stages in which disturbances caused by limited inputs can be compensated for without permanent structural changes. Webster, Waide & Patten (1975) therefore proposed replacing the term 'stability' with 'relative stability'. 'The argument that ecosystems are asymptotically stable focuses attention on the critical area of relative stability.' Their 'asymptotic stability' of an ecosystem is dependent upon the balance between all the structural elements of the system, its substance turnover and energy flow. This approach leads to an examination of the burdening capacity and elasticity of a system according to structural as well as energetic aspects. 'Resistance is related to the formation and maintenance of persistent ecosystem structure. Resilience results from the tendencies inherent in ecosystems for the erosion of such structures' (cf. among others Hall & Day 1977, May 1977, Patten 1974). In ecosystems with an abundance of species, single elements generally have a greater density of connections than in ecosystems with species paucity. It follows from this that ecosystems abundant in species can have a greater internal stability. With respect to external influences, however, they can be more labile or instable than ecosystems with a dearth of species. Stable ecosystems are able to overcome transformations caused by limited inputs, while labile ones depart permanently from their original states. The quantitative expression of a load is the degree of disturbance (Fränzle 1978). This indicates the measurable amount of the disturbance. Burdening therefore is the state up to which ecosystems can still overcome transformations forced on them. We do not consider it practical to limit the term 'burdening' to anthropogenic effects, because many of the 'loads' produced by man also occur naturally. Through an ecosystemic approach, the subjective concept of environment is replaced by a variable number of measurable parts and factors. One special case can be observed in key-species ecosystems. These are different in that their essential structural characteristics are decisively impressed by **one** element (e.g. animal or plant species, man). This means that the distribution of key-species ecosystems is identical to the area systems* of the respective species. If man were able, because of his technological abilities, among others, to take over the role of primary producer, consumer and decomposer simultaneously

– based for instance on a mastery of nuclear energies – he could maintain his cities (at least technologically) in a self-regulating manner. That man cannot live up to this expectation can be observed worldwide. A consequence of this is the burdening, alteration and destruction of the living systems of our earth. Since all systems are closely related for example to atmospheric happenings, the question as to how far man can change local systems without changing global processes irreversibly (among others Flohn 1977) is of increasing interest.

Systems whose ranges of stability are overstepped through the effects of exogenous impacts, can no longer recover (Ellenberg 1973). A system's ability to regenerate is decisively dependent upon the spatial effect a certain disturbance factor (e.g. pollutant; cf. Blau & Neely 1976) has on the density of connections and the type of webbing. During the last few years experimental biogeography was able to prove the existence of a close relationship between the size of an area and genetic structure (Calow 1977, Lack 1976, MacArthur 1972, MacArthur and Wilson 1971, Simberloff 1976, Wilson & Bossert 1973, among others). Numerous studies show that the rate of extinction in a system is contingent upon its size (Simberloff 1976). These findings led to the establishment of a catalogue of demands with respect to sizes of protected areas (Diamond 1975). They allow the conclusion to be drawn that in general only large areas can maintain their genetic structure rather constant over an extended period of time. The marked decrease in populations of numerous animal and plant species in West Germany can be attributed in a considerable number of cases to effects of area sizes.

2. Abiotic Factor Complexes and Ecosystems Behavior

Ecosystemic behavior is determined by endogenous element fluctuation and exogenous factors. This means that I can limit myself to one section of a system (e.g. element relations; factor behavior) in order to describe and control essential structural elements and energy flows. This is true. But before I can talk about 'ducks' (cf. Vester 1976), I have to know the system (among others Hall and Day 1977).

Therefore, it is necessary to know more about city climate, especially, however, about radiation conditions and their interaction with living systems and immission type (cf. among others Jost 1974, Junge 1972, Lamb 1977, Larson et al. 1975, Nguyen et al. 1974, Stern 1977, Umplis 1978). Equally important is knowledge about cycles of matter and the behavior of chemical elements (cf. among others Meyer 1977; further references in Müller 1977). Gaseous and partic-

---

* We understand an area system as an adaptive subsystem of the biosphere defined by the ecological valence, genetic variability and phylogeny of populations and the spatially and temporally changing effect of abiotic factors, that has ecological as well as phylogenetic functions and whose spatial dimension can be characterized through a three-dimensional distribution area of variable size and structure.

ulate emissions, radioactive substances and/or noise affect humans, animals and plants living in the cities and things as well. Interacting with spatial and climatic factors, they form the immission type (Hartkamp 1975, Fortak 1977, Kellogg 1977 among others). Photochemical processes and a great number of organic carbon compounds (cf. 1972 emission register of Cologne show that in establishing limits for air pollutants according to 'TA-Luft' and models of air pollution distribution (cf. among others Fortak 1977, Giebel 1977, 1978), obviously we are only working on the tip of the iceberg.

For every system, energy transfer and substance exchange are important parameters for describing and understanding its behavior. It seems te me that a fair amount of research has been done on many special climatological problems relevant to planning (among others behavior of fresh air; local wind systems) and their relation to immission type (e.g. how long pollutants stay in the atmosphere, modification and chemical composition of precipitation). However, not enough work has been done on the relationship of these problems to the most important ecosystems via radiation balance and dose effect relations. But especially these relationships are of great importance to the stability of the system (Müller 1977).

This does not underestimate the importance of other climatological and aerological research. For it was this kind of research that proved, for example, that the pollution of systems with sulfur compounds (e.g. SO_2, SO_4) can show very different distributional patterns. Among other things such research explained the low percentage of survival in organisms exposed at places that evidenced (at least according to gas analysis) equally small concentrations of SO_2.

3. Urban Ecosystems, 'Biotopes of Human Activity'

For pragmatic, methodological and historical reasons ecology as a science should only attempt to examine the interactions between living systems (organisms, man, populations, biocoenoses) and their environments, as far as they obey natural laws (for further information see Müller 1974; cf. also Schmithüsen 1976). Together with the theologian, Prof. Dr. U. Mann, (Müller & Mann 1978) I made clear that terms have to stay operable and that the present uncertainty of scientific prognosis (which in my opinion is caused by missing basic knowledge in many fields) cannot be allowed to become scientific anarchism, due to which scientifically based facts find themselves somewhere between ideology and/or profit.

Such a statement naturally has effects on an ecological approach to work in a city. In principal there is no rational argument that would hinder an application of the term ecosystem to a city. One has to take into consideration, however, that ecological work will only do justice to one particular level of information, namely the one that can be described with natural laws and not the comprehensive ecosystemic concept. Therefore, there are really at least five, in part very different approaches to ecological work:
a) 'city complex approach', which encompasses all the levels of human life and

activity and attempts to describe the city as a complex system with man as the center point. The development of this conception of a system can be traced from Park (1916) to Berry & Kasarda (1977), Burkhardt & Ittelson (1978) or Chapman (1977) and can also be found in the evaluation matrix of many city simulation models (cf. e.g. Schönebeck 1975, Wilson, Rees & Leigh 1977).

b) ecopsychological approaches, which begin to see everyday life as the 'biotope of human activity' (Boesch 1976), in which many, but not all things can be described in terms of natural laws (Eckensberger 1977, Rapoport 1977).

c) epidemiological approaches, dealing with the health of humans in urban areas (cf. among others Ehrlich et al. 1973, Rees et al. 1977, Rosmanith et al. 1975, Schlipköter 1977), based on distributional structures for example of illnesses (cancer, chronic bronchitis, cardiovascular diseases). In spite of some positive beginnings (e.g. Heidelberg, Münster), so far these projects have failed because the analyzed groups were not sufficiently well known (for instance social and genetic structure; cf. also Cherrett & Sagar 1977, Howe 1977).

d) energy analyses of cities and sites, which attempt to make prognoses for urban systems about energy flow and burdening capacities by way of in- and output analyses, biomass distribution, productivity and examination of radiation conditions (among others Döllekes 1977, Duvigneaud 1974, König 1977, Niehaus 1977, Voss 1977).

e) evaluation of urban ecosystems through decoding the ecological information content of the organisms living in them (summarizing literature among others in Ellenberg, Fränzle & Müller 1978, Müller 1978, Steubing et al. 1974, Sukopp et al. 1974). Besides registering all living systems in a city, bioindication (e.g. effects monitoring) and the analysis of food chains as pollution integrators are emphasized in regard to man. Pollutant accumulation, conversion and breakdown are also given special attention (Allen & Miller 1978, Booth & Ferrell 1977, Dougherty 1978, Khan 1977, Khan, Korte & Payne 1977, Kutz, Murphy & Strassman 1978, Matsumura 1977, Metcalf 1977, Rao 1978, among others).

I personally think that all of these approaches are important and I would not reject any one of them unless claims of preeminence were made for it at present. This is also true for cybernetical models, which certainly are only at the beginning of their development. In this connection we must not forget that 'there has to be a compromise between effort expended on a model study and the envisioned possibilities for results. Such a compromise does not allow for investigating all the interesting scientific details' (Voss 1977).

At the same time we are only at the beginning of our possibilities for ecological statements. Personally I believe that the analysis of epidemiological and energetic processes, the evaluation of city systems using bioindicators and the definition of the burdening of ecosystems through food chain analysis will help us progress most at the present, because right now it is necessary for us — more than ever before — to define the limitations for our decisions and actions and to recognize the fact that we ourselves and our systems are subject to biological laws.

4. Populations in the Wild and Area Evaluation

Seasonal and daily population dynamics and densities, in many cases perennial population cycles and/or species-specific population strategies determine the presence and therefore the verifiability of populations in their habitat. Without a knowledge of these population variables in space and time, we cannot use the informational content of living systems for the evaluation of areas and area surveys yield diametrical results (Berthold 1976, Diamond & May 1977). To put it 'sensitively', the 'duck' (Vester 1976, p. 38, 'The Duck Model') is not decisive; we have to know what species of duck or ducks live in a certain system in order to be able to estimate what is endogenously or exogenously controlled.

As an example I would like to use the present distribution in the Saar River of the hemerochorous water snail *Physa acuta*, which was unintentionally introduced. This species can live under beta to polysaprobic and obviously can also take high water temperatures. This caused problems with its classification according to the saprobic system (cf. among others Mauch 1977), but it enabled the species to colonize the Saar and its larger tributaries. In 1977 we found *Physa acuta* almost everywhere in the Saar, but there were considerable differences in its presence and population density during the course of the year, which fact can lead to misinterpretations of the water quality.

This example of *Physa* poses the question of the representativity of random samples, e.g. as proof of the water quality. Such samples will be required by the waste water law in West Germany from 1981 on. We all know that the effluents pouring into main drainage channels are more flexible than bureaucratic control systems. It therefore is not astonishing that abiotic parameters can fluctuate considerably according to pollutant, water body or time of day. This is true for thermal as well as heavy metal effluents. But a survey of the 221 invertebrates of the Saar — which we easily identified — and their classification according to substrate and saprobe groups provides us with an exact picture of the quality of the river's water. Once these relationships are known and for example the position of *Physa acuta* in the food web of the Saar has been clarified, *Physa* can be our 'indicator duck'. This requires total mobilization of effort for the task, while at the some time the methods of the causal analysis have to do justice to the complexity of the subject (Schmithüsen 1976). 'Without a solid biological foundation regional planning, ecological expertises, suggestions for measures of environmental protection can only be quackery: dangerous quackery, because simple and hasty solutions are so plausible' (Remmert 1978).

Today we know that besides the genetic structure, the biological abilities of populations are responsible for the dynamics of area systems. Knowing the biological abilities, however, means knowing the population growth that varies according to species, knowing the very different population regulators and knowing the population structure (age and sex ratios). It means a sure classification according to r- or K- strategies of their life style (Piank 1974), knowledge about their intra- and interspecific population behavior and about seasonal and daily rhythms as well as about their distribution. Populations are usually not distributed evenly within their area. This distribution may be influenced by chance, yet normally is closely correlated with the pattern of the habitat islands. Every organism and therefore every population chooses optimal habitats in a landscape first. Food supply, population density, territorial behavior, mates and pressure of enemies, among others, often result in a colonization of even suboptimal or pessimal habitats, something that my co-worker Dr. Hermann Ellenberg (1978) painstakingly proved for the roe deer in Central European cultural landscapes.

The importance that the interaction of population-genetic and ecological factors has for the decoding of the informational content of living systems for the evaluation of areas was elucidated by a number of recent scientific projects, at least for some of the major indicator species (synoptical treatment in Müller 1978).

5. Genetic Structure and Spatial Distribution

The intense preoccupation with present environmental problems, pollution limits and/or tests for compatability with the environment made us overlook that evolutionary processes take place at present in our cultural landscapes (and especially in our urban conglomerates), whose explanation is of the greatest importance for the evaluation of noxious substances. Of course each change in allele number already means evolution in the final analysis (Jain 1975, Lees et al. 1973, Kettlewell 1973, Maruyama 1977), but the isolation barriers and selection gradients that we built up accelerate this process. While we are watching, the 'ducks' ('The Duck Model'; Vester 1976) change. It proves wrong in many cases to talk about the ecological valence of a species, because this species may be composed of many populations with different ecological valences, which were only developed as a reaction to different burdening gradients. The subjects in question here go far beyond resistance problems with staphylococci and diptera strains, as well as coincident distributions of urban conglomerates and industry-menalistic lepidotera or coleoptera (among others *Biston betularia, Adalia bipunctata*). They point to questions that are partially connected with the differing bioaccumulation properties of pollutants and thus with the representativity of biological samples, as well as the validity of mutagenic, cancerogenic and teratogenic tests.

Beginning with the question why certain Coleoptera species, e.g. in Saarbrücken, are able to penetrate far into the urban area and why just the inner-city populations of these beetles prove to be more resistant in fumigation experiments, my co-worker, Dr. H. Steiniger (1978), examined the genetic structure of 7 Carabid populations in different sites. Using modern electrophoretic separation methods (disc electrophoresis with polyacrylamide gels), he was able to prove a correlation between allelepolymorphism and habitat diversities with 54 Carabid populations of urban and close-to-natural habitats (especially close with *Carabus violaceus, C. problematicus, Abax ater* and *Abax parallelus* as well as *Carabus nemoralis*).

215

But so far it was not possible to establish a causal relationship between isozyme-polymorphism and its selective advantages.

It is possible that the differing reactions of 'bioindicators' in various areas of our country are mostly due to genetic differences in populations, even though varying immission types and the related additive and synergistic behavior of pollutant combinations and climate types also play a role (among others Guderian 1977, Guderian, Krause & Kaiser 1977, Müller 1978).

6. Active Monitoring Programs and Residue Analyses

Genetic variability of wild populations and their daily and seasonal population fluctuations allow conclusions to be drawn about the quality of a particular area only if a high standard of information is achieved. Therefore, establishing active monitoring programs with 'standarized' organisms and substrates are necessary as an additional source of information (Arndt 1974, Guderian 1977, Müller 1975, Prinz 1975, Schönbeck & Van Haut 1974, Soraurer 1911, Steubing et al. 1974, 1976 and others).

In theory the necessity of effects monitoring programs was recognized early, because this technique shows the interaction of the ecologically important factors at one site especially well (among others climate type and immission components). For a large number of pollutants and pollutant combinations a dose-effect relation could be established and close correlations between specific pollutants, rates of accumulation in exposed organisms and their relative toxicity (e.g. as measured by the death rate of the exponates) were described. Most recent research also aims for correlations between emissions and reactions of exponates (among others Steubing et al. 1974, 1976, Rpu 1974, Scholl 1978).

Independent of which parameters of reaction are finally chosen as a guideline (among others CO_2 gas exchange; chlorophyll a; phenology) we can already say that active monitoring programs with e.g. Hypogymnia physodes or Lolium multiflorum are suited for practical tasks of area evaluation.

For the surrounding communities of Saarbrücken (Stadtverband) my co-worker W. Erhardt was able to establish significant correlations between air pollution and death rates of lichen thalli by means of a specific arrangement of exposed *Hypogymnia physodes*. The results of the 72 exposed series of lichens also pointed out air hygienic dangers in places where they were not recognized before (e.g. border region, Pflugscheid). Similarly to other exposition series, it also showed up here that the exact definition of the lichens at the beginning of the experiments, the duration of the exposition, treatment of the lichens and effects of the exposure site may have a strong influence on the reactions of the exponates. The question of representativity of an exposition site for a larger area therefore has to be established individually.

If the results of the exposition of Hypogymnia physodes for 140 days are compared with a map of the natural lichen vegetation that was constructed using

216

different methods and a map of forest growth damage, one can see direct correlations. This is also true for the results of residue analyses (Pb and Cd) of *Picea abies*, *Picea omorica* and *Picea pungens*, which were discovered by my co-worker G. Wagner.

Active monitoring programs for the Saar with limnic organisms were attempted (Müller & Schäfer 1976). After initial difficulties due to the choice of organisms, boxes and exposition sites, we were able to develop methods which allow a differentiation of the increase in biomass and the physiological condition of the exposed fishes, crawfish and molluscs as well as a correlation with the river's burdening pattern, resembling a mosaic.

Naturally death rates and pollutant accumulation can only indicate relative toxicity in regard to human populations and thus can only reflect a relative risk in certain sites (Kates 1978). But this is just as true, for instance, for all of our toxicological laboratory tests. Of course active monitoring programs can only provide information about one system factor complex (e.g. relative quality of the air or water), but I think this information is more important than the 12th annual series of SO_2 measurements in an area whose climatic type and emittent behavior I have known for a long time. One has to have the courage to rely on new parameters once a certain amount of a priori knowledge is available. In my opinion it is essential to establish the comparability of the different effects monitoring programs on a transnational scale as rapidly as possible and to come up with guidelines for their use in planning (e.g. VDJ).

7. Food Chain Analyses as Indicators for the Burdening of Areas

The populations and substrates linked in food-webs are integrators for the burden of areas and ecosystems. But area-related analyses of food-webs — at least in our country — have not progressed beyond their early stages. It is obvious that exclusive consideration of the benzpyrene content in the air (Hettche 1975) without an account of its derivatives in the human food chain (e.g. smoked herring; cf. Grimmer 1977) leads to area evaluation for which 'air' is a factor with unjustified decisiveness. This thought can easily be extended at least to recognize that all the substances we can measure have already been used as 'key substances' for burdening prognoses for air and water; the obvious question about transfer of substances within food chains, however, has not even been considered.

We know that the accumulation and breakdown of pollutants in individuals, populations and food chains can be accomplished in very different ways according, for example, to species, age and sex. While there are organisms whose pollutant accumulation is correlated with the concentration of the respective pollutant in their habitat, there are others that deposit selectively e.g. heavy metals and/or protect sensitive enzyme systems by binding the pollutants to proteins and thus reducing their toxicity. Consequently, if the area-specific food chains are not analyzed, residue analyses with birds (among others Conrad 1977, Joiris & Martens 1973, Koeman et al. 1973, Moore 1969, Prestt & Ratcliffe 1972), game animals (Brüggeman 1974, Drescher-Kaden 1976, among others), seals (Drescher et al. 1977) or invertebrates (among others Matsumura 1977) can only show how dangerous a pollutant is for a particular species but cannot indicate the burdening of an area. Goshawks (*Accipiter gentilis*), which mostly live on rabbits in one area, must necessarily have different PCB-contents than those of a different area where their main prey is wood pigeons (*Columba palumbus*).

For short food chains continuous analyses can be carried out; for food-webs, however, they can only be done sensibly via continuous ecosytem analysis and monitoring. Such a control can encompass essential environmental chemicals (among others Bevenue 1976, Gish & Christensen 1973, Martin & Coughtrey 1975, Müller 1978, Po-Yung et al. 1975) and radionucleids. Thus Miettinen (1976) for example examined and surveyed the whereabouts of the plutonium isotopes $^{239, 240}$Pu and ^{238}Pu over a period of ten years in a terrestrial food chain (lichens − reindeer − man) and an aquatic food chain (sediments − benthic invertebrate fauna − fishes). Similar analyses were carried out by Brisbin & Smith (1975) with American *Odocoileus virginianus* populations and with a soil arthropods food chain by Crossley, Duke & Waide (1975), in which cases ^{137}Cs was the main target (cf. among others also Dahlman, Francis & Tamura 1975). The influence of potassium on the whereabouts of cesium in *Rangifer tarandus* was studied by Holleman & Luick (1975) under controlled food intake conditions.

Transfer in food webs certainly has an influence on the chemical constitution and thus on the effectiveness of some pollutants. The methylation of mercury belongs in this category, as does the activation or the temporary inactivation of organic compounds. Basic problems with analyses, sometimes of the simplest kind, also show up when food-webs are analyzed. Reactions, which are wrongly attributed to certain substances (most of the time to those under examination) instead of to their derivatives and/or accompaning substances when isolated organisms are examined, can be interpreted correctly if food-webs are analyzed. As an example we can use the defoliant 2,4,5 trichlorophenoxyacetic acid:

$$Cl-\underset{Cl}{\overset{Cl}{\underset{\big|}{\bigcirc}}}-O-CH_2-COOR$$

In 1969 it was believed that 2,4,5 t was hazardous to small chickens (among other things the embryo died in the eggs of hens that were fed 2,4,5 t). Only later it was discovered that most likely neither the toxic effects nor the teratogenic properties could be attributed to 2,4,5 t itself but to an accompanying substance, namely terachlorodibenzo-p-diozin (TCDBD), which can also occur in 2,4,5 t in amounts of 0.1 − 30 ppm (Wegler 1977).

Food-webs are the basic structure for energy and matter transfer in ecosystems. This is why we think that the analysis of such food-webs is the main task we will have to work on in order to give a realistic answer to the question about the real pollution burden and endangerment for human populations. Especially research done in the last few years shows that it becomes more and more difficult to integrate all the multidimensional processes, that take place, for example, just within the immission type of a city, into a matrix for evaluating the situation for man. Improvement of already existing ('environmental compatibility tests') are being demanded more and more plainly. But these can only be used as relative reflections of a real risk to city life if the metabolism of this substance is analyzed in food chains of the respective system (among others Beavington 1975, Blau & Neely 1976, Jefferies & French 1976, Lagerwerff & Specht 1970, Williams & David 1976). Parallel to this we have to evaluate areas by means of effects monitoring programs with standardized organisms, the reactions or pollution residues of which yield information made transferable to human populations by epidemiological examinations (among others Guderian 1977, Guderian et al. 1977, Scholl 1978, Steubing et al. 1976).

Literature

Abbott, M. & Van Ness, H. 1976. Thermodynamik, Theorie und Anwendung. Mac Graw-Hill, New York.

Allen, J.R. & Miller, J.P. van 1978. Health Implications of 2,3,8,8 − Tetrachlorodibenzo-p-dioxin Exposure in Primates. In: RAO, Pentachlorophenol, 371-379, Plenum Press, New York.

Arndt, U. 1974. Langfristige Immissionswirkungen an ungeschütztem Nutzholz. *Staub* 34: 225-227.

Beavington, F. 1975. Heavy Metal Contamination of Vegetables and Soil in Domestic Gardens Around a Smelting Complex. *Environm. Polut.* 9: 221-217.

Berry, B. & Kasarda, J.D. 1977. Contemporary urban Ecology. Macmillan Publ., New York.

Berthold, P. 1976. Methoden der Bestandserfassung in der Ornithologie: Übersicht und kritische Betrachtung. *J. Ornithol.* 117 (1): 1-69.

Bevenue, A. 1976. The 'bioconcentration' aspects of DDT in the environment. Residue Rev. 61: 37-112, Springer, Heidelberg.

Blau, G. & Neely W. 1976. Mathematical Model Building with an Application to determine the Distribution of Dursban Insecticide added to a Simulated Ecosystem. *Adv. Ecol. Research* 9: 133-163.

Boesch, E. 1976. Psychopathologie des Alltags. Zur Ökopsychologie des Handelns und seiner Störungen. Huber, Bern.

Booth, G.M. & Ferrell, D. 1977. Degradation of Dimilin[®] by Aquatic Foodwebs. In: KHAN, Pesticides..., 221-243, Plenum Press, New York.

Brisbin, I.L. & M.H. Smith 1975. Radiocesium concentrations in Whole-Body Homogenates and several Body Compartments of naturally contaminated White-Tailed Deer. Mineral Cycling in Southeastern Ecosystems, Springfield, Virginia.

Brüggemann, J. 1974. Pestizid- und PCB-Rückstände in Organen von Wildtieren als Indikatoren für Umweltkontaminationen. *Z. Jagdwiss.* 20: 70-74.

Buchwald, K. 1977. Landschaftsplanung als Beitrag zur Standortbeurteilung und Standortfindung für Energieanlagen aus ökologischer und gestalterischer Sicht. In: Energie und Umwelt 58-61, Düsseldorf.

Burkhardt, D. & Ittelson, W. 1978: Environmental assessment of Socioeconomic systems. Plenum Press, New York.

Calow, P. 1977. Ecology, Evolution and Energetics: A Study in Metabolic Adaptation. In: *Ecological Research* 10: 1-62.

Chapman, G. 1977. Human and Environmental Systems. A Geographer's Appraisal. Acad. Press, New York.

Cherrett, J.M. & Sagar, G.R. 1977. Origins of Pest, Parasite, Disease and Weed Problems. Blackwell Publ., Oxford.

Conrad, B. 1976. Die Giftbelastung der Vogelwelt Deutschlands. – Vogelkundliche Bibliothek, Bd. 5, Kilda-Verlag, Greven. 68 S.

Crossley, D., Duke, K. & Waide, J. 1975. Fallout Cesium-137 and Mineral-Element Distribution in Food chains of Granitic-Outcrop ecosystems. In: Mineral Cycling in Southeastern Ecosystems, 580-587, ERDA.

Cumberland, J.D. 1977. Boundary condition and influence on area planning of the power generating industries. In: Energie und Umwelt, 43-45, Düsseldorf.

Dahlman, R., Francis, C. & Tamura, F. 1975. Radiocesium cycling in vegetation and soil. In: Mineral Cycling in Southeastern Ecosystems, 462-481, ERDA.

Diamond, J.M. 1975. The island dilemma: Lessons of modern biogeographic studies for the design of natural reserves. *Biol. Conserv.* 7: 129-146, Applied Science, Publ., Great Britain.

Diamond, J.M. & May, R.M. 1977. Species turnover rates on islands: dependence on census interval. *Science* 197: 226-270.

Dölleke s, H. 1977. Ein multisektorales Energie- und Umweltplanungsmodell. In: König, CH. 1977, Energiemodelle..., Birkhäuser Verl., Basel.

Dougherty, R. 1978. Human Exposure to Pentachlorophenol. In: RAO, Pentachlorophenol, 351-361, Plenum Press, New York.

Drescher, H.E., Harms, U. & Huschenbeth, E. 1977. Organochlorines and Heavy Metals in the Harbour Seal Phoca vitulina from the German North Sea Coast. *Marine Biology* 41: 99-106.

Drescher-Kaden, U. 1976. Nationale und internationale Forschungsaktivitäten und Ergebnisse auf dem Gebiet der Nutzung freilebender Tierarten als Indikatoren für die Belastung der Umwelt – insbesondere des Menschen – durch Umweltchemikalien. Bundesministerium für Jugend, Familie und Gesundheit, Bonn-Bad-Godesberg.

Duvigneaud, P. 1974. L'écosystème 'URBS'. *Mém. Soc. roy. Bot. Belg.* 6: 5-35.

Ellenberg, Heinz 1973. Die Ökosysteme der Erde, Versuch einer Klassifikation der Ökosysteme nach funktionalen Gesichtspunkten. In: Ökosystemforschung, Springer, Heidelberg.

Ellenberg, Heinz, Fränzle, O. & Müller, P. 1978. Ökosystemforschung im Hinblick auf Umweltpolitik und Entwicklungsplanung. Denkschrift erstellt im Auftrag des Bundesministers des Innern vertreten durch das Umweltbundesamt UBA, Berlin.

Ellenberg, Hermann 1978. Zur Populationsökologie des Rehes (Capreolus capreolus L., Cervidae) in Mitteleuropa. – Beiheft zur Zeitschrift für Zoologie 'Spixiana', Zool. Staatssammlung, München, ca. 200 S. im Druck.

Eckensberger, L. 1977. Die ökologische Perspektive in der Entwicklungspsychologie: Herausforderung oder Bedrohung? Symp. Entwicklung in ökologischer Sicht, Konstanz.

Ehrlich, P., Ehrlich, A. & Holdren, J.P. 1973. Human Ecology. Problems and Solutions. Freeman, San Francisco.

Flohn, H. 1977. Großräumige Beeinflussung des Klimas durch menschlichen Eingriff? Historischer Überblick und künftige Aussichten. In: Energie und Umwelt, 110-113, Düsseldorf.

Fortak, H.G. 1977. Beeinflussung des Lokalklimas durch große Energieerzeugungs- und Verbraucherzentren. In: Energie und Umwelt 113-117, Düsseldorf.

Fränzle, O. 1978. Die Struktur und Belastbarkeit von Ökosystemen. 41. Dtsch. Geographentag Mainz, 469-485, Steiner, Wiesbaden.

Giebel, J. 1977. Untersuchungen zur Simulation der Immissionsbelastung durch Schwefeldioxid. in der Umgebung von Ballungsräumen. Schriftenr. Landesanst. *Immissionssch. NRW* 42: 17-31, Essen.

Giebel, J. 1978. Berücksichtigung der trockenen Ablagerung am Boden in Abhängigkeit von der effektiven Quellhöhe bei einem Gausschen Ausbreitungsmodell. Schriftenr. Landesanst. *Immissionssch. NRW* 43: 26-35, Essen.

Gish, C.D. & Christensen, R.E. 1973. Cadmium, Nickel, Lead and Zinc in Earthworms from Roadside. Oil. *Environ. Sci. Technol.* 7: 1060-1062.

Guderian, R. 1977. Air Pollution. Phytotoxicity of Acidic Gases and its Signifiance in Air Pollution Control. Ecological Stud. 22, Springer, Heidelberg.

Guderian, R., Krause, G. & Kaiser, H. 1977. Untersuchung zur Kombinationswirkung von Schwefeldioxid und schwermetellhaltigen Stäuben auf Pflanzen. Schiftenr. Landesanstalt für Immissionsschutz NRW 40: 23-30, Essen.

Hall, Ch. & Day, J. 1977. Ecosystem Modeling in Theory and Practice: An Introduction with Case Histories. Wiley, New York.

Hartkamp, H. 1975. Untersuchungen zur Immissionsstruktur einer Großstadt. Projekt 'Groß-stadtluft'. *Schriftenreihe der LIB* 83: 30-38.

Hettche, O. 1975. Zum Problem eines Immissionsgrenzwertes für Benzo (A) Pyren. *Umwelthygiene* 2: 46-50.

Holleman, D.F. & J.R. Luick 1975. Relationship between Potassium intake and Radiocesium retention in the Reindeer. Mineral Cycling in Southeastern Ecosystems, Springfield Virginia.

Howe, G.M. 1977. A World Geography of Human Diseases. Acad. Press, London.

Jain, S.K. 1975. Patterns of Survival and Microevolution in Plant Population.

Jefferies, D.J. & French, M.C. 1976. Mercury, Cadmium, Zinc, Copper, and Organochlorine. Insecticide Levels in Small Mammals Trapped in a Wheat Field. *Environm. Pollut.* 10: 175-182.

Joiris, C. & Martens, P. 1973. Teneur en pesticides organochlores d'oeufs de rapaces recoltes en Belgique en 1971. *Aves* 10: 153-160.

Jost, D. 1974. Aerological Studies on the Atmospheric Sulfur Budget. *Tellus* 26.

Junge, C.E. 1972. The Cycle of Atmospheric gases-Natural and Man Made. *Quart. J. Royal Met. Soc.* 98.

Kates, R.W. 1978. Risk Assessment of environmental Hazard. SCOPE 8, Preface, Salisburg.

Kellogg, W. 1977. Emissions and the Climate. In: Energie und Umwelt 117-123, Düsseldorf.

Kettlewell, H. 1973. The evolution of melanism. Oxford.

Khan, M.A. 1977. Pesticides in Aquatic Environments. Plenum Press, New York.

Khan, M., Korte, F. & Payne, J. 1977. Metabolism of Pesticides by aquatic Animals. In: Khan, Pesticides... p. 191-220, Plenum Press, New York.

Koeman, J.H. et al. 1973. Effects of PCB and DDE in cormorans and evaluation of PCB residues from an experimental study. *J. Reprod. Fert., Suppl.* 19: 353-364.

König, Ch. 1977. Energiemodelle für die Bundesrepublik Deutschland. Birkhäuser Verl., Basel.

Kutz, F.W. Murphy, R. & Strassman, S.C. 1978. Survey of Pesticide Residues and their Metabolites in Urine from the general Population. In: RAO, Pentachlorphenol, 363-369, Plenum Press, New York.

Lack, D. 1976. Island Biology, illustrated by the land birds of Jamaica. Studies in Ecology 3, pp. 445, Blackwell Scient. Publ. Oxford.

Lagerwerff, J.V. & Specht, A.W. 1970. Contamination of Roodside Soil and Vegetation with Cadmium, Nickel, Lead and Zinc. *Environ. Sci. Techn.* 4 583-586.

Lamb, H.H. 1977. Climate. Present, Past and Future. Methuen, London.

Larson, T. et al. 1975. The influence of a Sulfur Dioxide Point Source on the Rain Chemistry of a Single Storm in the Puget Sound Region. *Water Air Soil Pollut.* 4.

Lees, D.R., Creed, E.R. & Duckett, J.G. 1973. Athmospheric pollution and industrial melanism. *Hèredity* 30: 227-232.

MacArthur, R.H. 1972. Geographical Ecology, Patterns in the Distribution of Species, Harper & Row Publ., New York.

MacArthur, R.H. & Wilson, E.O. 1971. Biogeographie der Inseln, Verl. Goldmann, München.

Mann, U. & Müller, P. 1978. Bewertungsmatrix und Lösungsstrategien für gegenwärtige ökologische und ökonomische Probleme. Ergebnisse eines Dialoges zwischen Theologen und Biogeographen in Saarbrücken; 1974-1977. Mitt. *Schwerpunkt Biogeographie Univ. Saarlandes*, 10: 1-15, Saarbrücken.

Margalef, R. 1975. Perspectives in ecological theory. Chicago, London.

Martin, M.H. & Coughtrey, P.J. 1975. Preliminary Observations on the Levels of Cadmium in a Contaminated Environment. *Chemosphere* 3: 155-160.

Maruyama, T. 1977. Stochastic Problems in Population Genetics. Springer, Heidelberg.

Matsumura, F. 1977. Absorption, accumulation and elimination of Pesticides by aquatic organisms. In: Khan, Pesticides..., 77-125, Plenum Press, New York.

Mauch, E. 1976. Leitformen der Saprobität für die biologische Gewässeranalyse. Cour. Forsch. Senckenberg 21.

May, R.M. 1972. Will a Large Complex System be Stable? *Nature* 238: 413-414.

May, R.M. 1973. Stability and Complexity in Model Ecosystems. Princeton University Press, Princeton, N.Y.

May, R.M. 1977. Thresholds and breakpoints in ecosystems with a multiplicity of stable states. *Nature* 269: 471-477.

Metcalf, R. 1977. Modell Ecosystem Studies of Biocentration and Biodegradation of pesticides. In: Khan, Pesticides..., 127-144, Plenum Press, New York.

Meyer, B. 1977. Sulfur, Energy, and Environment. Elsev. Publ., Amsterdam.

Miettinen, J.K. 1976. Plutonium Foodchains. Helsinki.

Moore, N.W. 1969. Sublethal effects of organochlorine insecticides in the Laboratory and the field. Meded. Fakult. *Landbouw-wetenschappen Gent* 34: 408-412.

Müller, P. 1974. La structuration de l'environnement naturel dans les régions de concentration urbaine, Bull. des Recherches agronomiques de Gembloux, Vol. extraordinaire édité à l'occasion de la semaine d'étude agriculture et environnement, P. 742-761, Gembloux.

Müller, P. 1974. Was ist Ökologie? *Geoforum* 18.

Müller, P. 1975. Ökologische Kriterien für die Raum- und Stadtplanung, Umwelt-Saar 1974, 6-51, Homburg.

Müller, P. 1976. Tiere als Belastungsindikatoren und ökologische Kriterien. *Daten und Dokumente zum Umweltschutz* 19: 153-171, *Stuttgart*.

Müller, P. 1977. Die Belastbarkeit von Ökosystemen. In: Energie und Umwelt, Envitec, Düsseldorf.

Müller, P. 1977. Biogeographie und Raumbewertung. Wiss. Buchgesellsch. Darmstadt.

Müller, P. 1977. Tiergeographie. Teubner, Stuttgart.

Müller, P. 1978. Erfassung von Arealsystemen – eine Grundlage für die Raumbewertung. Beitrag zur Aufschlüsselung des Informationsgehaltes von Tierarealen für die Darstellung der Umweltsituation der Bundesrepublik Deutschland. In: Schriftenr. Landesanstalt für Umweltschutz BW, Karlsruhe.

Müller, P. 1978. Biogeographie, UTB, Stuttgart (im Druck).

Müller, P. 1978. Arealsysteme und Biogeographie. Ulmer, Stuttgart (im Druck).

Müller, P. & Schäfer, A. 1976. Diversitätsuntersuchungen und Expositionstests in der mittleren Saar. *Forum Umwelthygiene* 2: 43-46.

Nguyen ba Cuong, Bonsang, B. & Lambert, G. (1974): The Atmospheric Concentration of Sulfur Dioxide and Sulfate Aerosols over Antarctic, Subantarctic Areas and Oceans, *Tellus* 26.

Niehaus, F. 1977. Computersimulation langfristiger Umweltbelastung durch Energieerzeugung. Kohlendioxyd, Tritium und Radio-Kohlenstoff. Birkhäuser Verl., Basel.

Park, R. 1916. The City: Suggestion for the Investigation of Human Behavior in an Urban Environment. *Amer. Journ. Sociology* 20: 577-612.

Patten, B.C. 1974. The Zero State and Ecosystems Stability. Proc. First Internat. Congr. Ecology. Centre for Agricult. Publ. and Documentation, Wageningen.

Pianka, E.R. 1974. Evolutionary Ecology. Harper & Row Publ., New York.

Po-Yung, L., R.L. Metcalf, R. Furman, R. Vogel & J. Hasset 1975. Model Ecosystem Studies of Lead and Cadmium and of Urban Sewage Sludge Containing this Elements. *J. Environ. Qual.* 4.

Prestt, J. & Ratcliffe, D.A. 1972. Effects of organochlorine insecticides on European birdlife. Proc. XV th intern. ornith. Congr. 1970: 486-513.

Prinz, B. 1975. Immissionswirkungskataster in Nordrhein-Westfalen als Planungskriterium. Umwelt-Saar 1974.

Rao, K.R. 1978. Pentachlorphenol. Chemistry, Pharmacology, and Environmental Toxicology. Plenum Press, New York.

Rapoport, A. 1977. Human Aspects of Urban Form. Towards a Man-Environment Approach to Urban Form and Design. Pergamon Press, Oxford.

Rees, P.H., Smith, A.P. & King, J.R. 1977. Population models of Cities and Regions. John Wiley, New York.

Remmert, H. 1978. Ökologie. Ein Lehrbuch. Springer, Heidelberg.

Rosmanith, J., Schröder, A., Einbrodt, H.J. & Ehm, W. 1975. Untersuchungen an Kindern aus einem mit Blei und Zink belasteten Industriegebiet. *Umwelthygiene* 9: 266-271.

Rpu = Regionale Planungsgemeinschaft Untermain 1974. Lufthygienisch-meteorologische Modelluntersuchung in der Region Untermain. 5. Jahresbericht. Frankfurt.

Schlipköter, H.-W. 1977. Bewertung der Grenzwerte von Immissionen aus medizinischer Sicht. In: Energie und Umwelt 100-105, Düsseldorf.

Schmithüsen, J. 1976. Allgemeine Geosynergetik. Grundlagen der Landschaftskunde, De Gruyter, Berlin.

Scholl, G. 1978. Vergleich zwischen Emissionsrate und Wirkdosis von Fluorid in standardisierten Graskulturen im Bereich eines Aluminiumwerkes. Schriftenr. *Landesanst. Immissionsschutz NRW* 43: 75-89, Essen.

Schönbeck, H. & Van Haut, H. 1974. Methoden zur Erstellung eines Wirkungskatasters für Luftverunreinigungen durch pflanzliche Indikatoren. Verhdl. Ges. Ökologie, 435-445, Saarbrücken. Verl. Junk, The Hague.

Schönebeck, C. 1975. Der Beitrag komplexer Stadtsimulationsmodelle (vom Forrester-Typ) zur Analyse und Prognose großstädtischer Systeme. Birkhäuser, Basel.

Simberloff, D.S. 1976. Trophic structure determination and equilibrium in an arthropod community. *Ecology* 57: 395-398.

Sorauer, P. 1911. Die makroskopische Analyse rauchgeschädigter Pflanzen. Samml. Abhdl. Abgase und Rauchschäden 7.

Steiniger, H. 1978. Genetische Variabilität bei Carabiden-Populationen inner- und außerstädtischer Standorte. Dissertation, Biogeographie, Saarbrücken.

Stern, A. 1977. Air Pollution. 3. Ed., Vol. 5, Air Quality Management. Acad. Press, New York.

Steubing, L., Kirschbaum, U. & Gwinner, M. 1976. Nachweis von Fluorimmissionen durch Bioindikatoren. *Angew. Botanik* 50: 169-185.

Sukopp, H. et al. 1974. Ökologische Charakteristik von Großstädten, besonders anthropogene Veränderungen von Klima, Boden und Vegetation. *TUB* 6 (4): 469-488.

Thoss, R. 1977. Möglichkeiten einer optimalen Standortverteilung von Energieversorgungsanlagen unter Berücksichtigung gesamtwirtschaftlicher Aspekte. Energie und Umwelt, 45-49, Düsseldorf.

Umplis, 1978. Verzeichnis rechnergestützter Umweltmodelle. E. Schmidt Verl., Berlin.

Vester, F. 1976. Ballungsgebiete in der Krise. Eine Anleitung zum Verstehen und Planen menschlicher Lebensräume mit Hilfe der Biokybernetik. Biering, München.

Voss, A. 1977. Ansätze zur Gesamtanalyse des Systems Mensch-Energie-Umwelt. Eine dynamische Computersimulation. Birkhäuser, Basel.

Webster, J., Waide, J. & Patten, B. 1975. Nutrient recycling and the stability of Ecosystems. Mineral Cycling in Southeastern Ecosystems. ERDA Sympos. Ser., U.S. Energy Research and Development Administration, Springfield.

Wegler, R. 1977. Herbizide. In: Chemie der Pflanzenschutz- und Schädlingsbekämfungsmittel, 5. Springer, Heidelberg.

Williams, C.H. & David, D.J. 1976. The Accumulation in Soil of Cadmium Residues from Phosphate Fertilizers and their Effect on the Cadmium Content of Plants. *Soil Science* 121: 86-93.

Wilson, A.G., Rees, P. & Leigh, C. 1977. Models of Cities and Regions. Theoretical and Empirical Developments. John Wiley, New York.

Wilson, E. & Bossert, W. 1973. Einführung in die Populationsbiologie. Springer, Heidelberg.

Adress of the author:
Prof. Dr. er. nat. Paul Müller
Lehrstuhl für Biogeographie, Geographisches Institut
Universität des Saarlandes
66 Saarbrücken, FRG